An Atlas of Cytopathology of the Head and Neck

Dedicated to my parents Claud and Sue Warner and to my siblings Suzanne, Sylvia, Tom, Harry and Lucy

An Atlas of Cytopathology of the Head and Neck

with clinical and histological correlations

Geneviève Warner Learmonth MB, BCh, BAO(NUI), FFPath(Anat)SA, MIAC
Honorary Senior Consultant Lecturer in Cytopathology and Anatomical Pathology, University of Cape Town and Groote Schuur Hospital, Cape Town, South Africa; formerly Head of the Cytopathology Unit, Department of Anatomical Pathology, Groote Schuur Hospital and University of Cape Town, South Africa

ARNOLD
A member of the Hodder Headline Group
LONDON · SYDNEY · AUCKLAND
Co-published in the USA by Oxford University Press Inc., New York

First published in Great Britain in 1998 by
Arnold, a member of the Hodder Headline Group,
338 Euston Road, London NW1 3BH

http://www.arnoldpublishers.com

Co-published in the United States of America by
Oxford University Press Inc.,
198 Madison Avenue, New York, NY10016
Oxford is a registered trademark of Oxford University Press

Whilst the advice and information in this book are believed to be true and accurate at the date of going to press, neither the authors nor the publisher can accept any legal responsibility or liability for any errors or omissions that may be made. In particular (but without limiting the generality of the preceding disclaimer) every effort has been made to check drug dosages; however, it is still possible that errors have been missed. Furthermore, dosage schedules are constantly being revised and new side-effects recognized. For these reasons the reader is strongly urged to consult the drug companies' printed instructions before administering any of the drugs recommended in this book..

British Library Cataloguing in Publication Data
A catalogue record for this book is available from the British Library

Library of Congress Cataloging-in-Publication Data
A catalog record for this book is available from the Library of Congress

ISBN 0 340 74103–1

1 2 3 4 5 6 7 8 9 10

Typeset in 10½/12 pt Palatino by Typestudy, Scarborough, North Yorkshire
Printed and bound in China

What do you think about this book? Or any other Arnold title?
Please send your comments to feedback.arnold@hodder.co.uk

Contents

Preface

During my years in the practice of cytopathology – both in routine screening and in 'on-site' diagnosis – I have noted that the books most frequently present at the elbow of the microscopist were usually atlases with many photomicrographs, short descriptive legends, a minimum of text and a few key references. A comprehensive ready reference-type atlas solely dedicated to cytopathology of the head and neck was not available, although several textbooks contain excellent chapters on salivary glands, neck nodes etc. A single atlas for 'bench reference' was needed to assist those involved in 'on-site' diagnosis in busy, clinically orientated cytopathology laboratories. Difficult, controversial and unusual cases can be researched in the medical literature and in larger textbooks on selected pathology topics.

This *Atlas of Cytopathology of the Head and Neck* is intended primarily for diagnostic microscopists, pathologists and clinicians involved in the diagnosis of oncological disease. The clinical and histological correlation will be of interest to all those concerned with the fine structural details of the cell and in the clinically relevant biological behaviour that cellular morphological changes reflect. Techniques for collecting specimens are not discussed in this atlas as there are many excellent and detailed publications that cover these.

Cytopathology is an intensely demanding discipline and skill; it is also an art, and a very visual art at that. It is not surprising that many aspiring artists are to be found masquerading as cytopathologists! Therefore I have endeavoured to present photomicrographs that are visually stimulating, yet meaningful in their depiction of the structural details of cells, as related to their clinically relevant behaviour. Routine Papanicolaou stain which produces a rainbow of colours, was used in the majority of cases in the author's clinical and academic unit (see Introduction). Occasionally special stains and immunocytochemical stains were used to solve difficult clinical problems.

The photography of microscopic slides is challenging. In a busy laboratory it is often difficult to find time to shoot photomicrographs. I kept my camera constantly at the ready so that I could photograph as I reported, without interrupting the work flow. Because the cell clusters and microbiopsies encountered on cytological preparations are three-dimensional, it can be difficult to create meaningful photographs while depicting fine cytological detail. Furthermore, as there may be up to 300 000 cells on a cytology slide, and the significantly abnormal cells may be at opposite ends, it is challenging to capture cohesive and instructive images. Sometimes this challenge was overwhelming and I failed to produce the perfect photomicrographic vignette. However, I hope that what I have presented will be evaluated with these points in mind.

The text and the legends have been designed to provide additional background information and teaching discussion. Selected cases have been included to reinforce a particular diagnostic problem.

Diagnostic cytopathology cannot be practised in a vacuum. I have attempted to emphasize the critical importance of interpreting the cellular changes while keeping the clinical findings and/or possibilities constantly in mind. I have found it extremely useful to keep a mental list of differential diagnoses when confronted with a cytopathological problem, and to keep an open mind about this list. In this way I believe that diagnostic cytopathologists can offer the optimum service to patients.

Cytopathology of the head and neck is important because:

- most FNA specimens from metastatic lesions are from neck nodes
- the majority of lymphoproliferative lesions initially present in the neck
- tuberculous lesions, which are easily accessible, are generally in the neck
- cancerous lesions of the skin (including melanoma) are commoner in the head and neck than in other areas of the body
- oral and nasopharyngeal cancer is common in South Africa, and a cheap diagnostic modality was needed to assist in the management of the innumerable patients attending ENT outpatient and primary health care clinics
- in my practice the number of specimens from the

head and neck began to outnumber those from the breast or lung.

Exfoliative cytology of the oral cavity, nasopharynx, paranasal sinuses and nose has not been widely used in recent times, although judging by the number of publications it was extensively discussed in the 1960s and 1970s. I initially studied cytobrushings from the oral cavity and nasopharynx to assess their potential use in the triage management of outpatients. Gradually my clinical colleagues became convinced of the diagnostic value of oral cytology and ultimately used this diagnostic tool in the triple assessment of patients with clinical lesions.

As every experienced diagnostic cytopathologist can appreciate, 'to be uncertain is uncomfortable; but to be certain is ridiculous' (Goethe). Nothing is certain in cells and nuclei.

GWL
Cape Town, 1998

Acknowledgements

I wish to acknowledge the support and encouragement of all my colleagues, family and friends around the world. They are too numerous to mention individually. However, I am especially grateful to the late Professor C. J. Uys of Cape Town for encouraging me with my postgraduate studies; to Dr Anne Linder for introducing me to the *clinical approach* to cytopathology; and to Dr Moira Durcan for constantly spurring me on: indeed, she initiated my efforts to compile this atlas. Dr Malcolm Hayes (Vancouver) encouraged me to persist with non-gynaecological and fine-needle aspiration cytopathology, especially the 'on-site' clinics. Dr Bryan Knight has sacrificed many hours of his time in offering valuable criticism of the illustrations.

Without the enthusiastic support of the cytotechnologists, the laboratory staff and the clerical staff of the Yvonne Parfitt Cytopathology Unit, there would not have been such beautifully prepared material for me to work on. To the staff who accepted the responsibility of the on-site clinics I am particularly grateful. The meticulous computer and clerical skills and record-keeping of Dolores Kleinsmith and her team, Jane Beck and Sally Kilbey, enabled me to assemble correlative histological material and clinical details.

I wish to thank the clinicians from Groote Schuur Hospital who supported and trusted me, and who sent me such well-prepared and interesting specimens, supported by clinical documentation: in particular, Professor Sean Sellars and Dr John Duff of the Ear, Nose and Throat Department; from General Surgery Professors Dent and Immelman; from the Ultrasound Department Dr Hillier Goodman; from the Radiotherapy Department Professors Dudley Werner and Raymond Abratt, Drs Clare Stannard, Anne Hacking and Anne Gudgeon – a fine aspirator! The physicians from the Respiratory Unit provided valuable material and information: Professors Benatar, Bateman and Willcox.

Much of my work was supported by the National Cancer Association of South Africa, the Becker Trust for the Elderly, and the Leon Goldstein Trust (Cape Town).

I learnt so much from my pupils' constant questions. I thank them for enduring my 'deep-immersion' teaching technique; indeed, some learnt by optical osmosis!

Without my patients this atlas could not have been compiled. This book in reality belongs to them.

Michael Wyeth and David Levin from the Department of Medical Photography carried on producing excellent photographic work with very limited resources, despite the academic and economic sanctions.

Last but not least, I would like to express my appreciation to my little family Ian, Despina, Suzanne and Sean, for their forbearance during the preparation of the illustrations and manuscript. Without Sean's sense of humour, on-the-spot assistance and computer literacy, I could not have completed this work.

Introduction

Diagnosis of malignancy is generally regarded as the major objective of clinical cytopathological diagnosis. However, recognition of benign processes – infective, inflammatory, degenerative or neoplastic – is of equal importance in the triage management and follow-up of patients.

As in all other areas of medicine, a holistic approach to the clinical problem is of paramount importance in clinical cytopathology. A sound knowledge of basic pathological processes and an in-depth knowledge of systemic anatomical pathology is a prerequisite for the provision of a reliable diagnostic cytopathology service.

With experience, the microscopist begins to appreciate the limitations and advantages of cytopathology. This experience can only be acquired at the microscope, in the clinic and at clinicopathological meetings.

The choice of stain in cytological services is controversial and a popular subject for debate. In a South African laboratory processing approximately 90 000 specimens per year, and labouring under severe economic and political sanctions and academic boycotts, it was necessary to be innovative and to cut costs. It was also necessary to be realistic. It was therefore decided to have one stain for all specimens – Papanicolaou's multichrome 'rainbow' stain. This is first and foremost a good haematoxylin stain. The eosin is alcoholic for increased structural clarity and transparency. The two counterstains (also alcoholic) were orange G and light green, which yield a wide range of colour reactions covering the entire visible spectrum from 400 to 700 nm.

In using one stain for all specimens:

- uniformity of specimen preparation by healthcare workers from all types of clinics was maintained
- work flow in the laboratory preparation room was smooth
- automated slide-staining was facilitated
- teaching of cytotechnologists was facilitated, as they had all been introduced to cytology with the screening of Papanicolaou-stained cervical smears. Equally, it was easier for these students to look at correlative H&E sections
- teaching of registrars in histopathology was facilitated, as their store of ability and knowledge built up on H&E-stained sections was transferable to the Papanicolaou-stained specimens
- coping with the nuances of air-drying artefacts was obviated.

Cytopathological diagnosis of disease depends on the assessment of deviation from the norm. The relevant morphological features of unregulated growth are seen principally in the nucleus, whereas defects of cell maturation are manifested by changes in cytoplasmic features. Abnormal cell function can often be demonstrated by special staining techniques.

BACKGROUND EVALUATION

The appreciation of background is of great importance in the total evaluation of the biological behaviour of the cellular material on a smear. The type of inflammatory exudate, the presence of blood, the extent and nature of necrosis, and the identification of organisms and contaminants all add to the eventual recognition of the underlying biological processes. The subconscious habitual examination of the background material often prevents erroneous diagnoses being made.

If a fine-needle aspiration (FNA) specimen contains eosinophils and carcinoma, the cytologist should not ignore the former because the specimen was submitted for the diagnosis of the latter. The presence of eosinophils may be indicative of fungal infection, which may suggest diabetes, or it may be associated with immunosuppression (or both), so that several biological processes may ultimately be recognized and treated.

Pigments

As discussed in several chapters in this book, the recognition of pigment and apparent 'dust' contaminants can be subtle and puzzling; however, they should never be ignored, as the explanation of their nature and their presence may lead to a correct, if unexpected, diagnosis.

Necrosis

A necrotic background indicates cell death from whatever cause. Caseous necrosis is a peculiar curd-like necrosis which occurs in the centre of tuberculous lesions. It appears as amorphous chromatic material (eosinophilic or basophilic), with necrotic nuclear debris scattered throughout. Wisps of 'comma'-shaped basophilic material are often seen.

Liquefying necrosis, which occurs frequently in cysts, produces a watery granular background in which degenerate cells and debris are present.

Lymphoid cells

In reactive states large immature lymphocytes may lead to an erroneous diagnosis of lymphoma by the unwary; however, the proper identification of tingible-body macrophages among these will point to their origin in reactive follicles. Their regular nuclear structure is also helpful in determining their benign origin. Widespread 'individual cell necrosis' is peculiar to rapidly progressive lymphoma.

Cellularity of the specimen

A very cellular specimen usually indicates a representative specimen. Sometimes marked cellularity may be the only indicator of malignancy in very well differentiated adenocarcinomas. A scanty cellular yield in FNA specimens may indicate a geographical miss, poor aspiration technique or aspiration of a stroma-rich lesion.

The cohesiveness of cells is of utmost importance in assessing malignancy in FNA specimens. Malignant and dysplastic cells lose their intercellular cohesion, lying in small groups or clusters with apparent loose cohesion, or they lie singly (lymphoid cells always lie singly). Cell-to-cell variability is quite striking in most malignant groups. Moulding of the nuclei of adjacent cells is seen in rapidly growing tissues, whether they are reactive or neoplastic; however, overlapping of nuclei is usually a reliable indicator of neoplasia in sheets of cells.

Papillary groups, numerous acinar arrangements, leashes of capillaries, rosette formation, globules of hyaline material and sheets of fibromyxoid material are all abnormal findings and should be assessed within the limits of the anatomical site.

NUCLEI

Size

The amount of cytoplasm and nucleus in a cell is expressed as the nucleocytoplasmic (N/C) ratio. This is important in the assessment of reactive changes and of malignancy. In malignant cells the nuclear enlargement is disproportionately great and is erratic (the nuclei vary in size and shape), whereas nuclear enlargement in benign processes is uniform and the nuclear shape is normal. In benign cells which are adjacent to a malignant process the reactive changes may be so extreme as to cause confusion. These changes may be ascribed to the production of tumour growth factor by the malignant neoplasm.

Staining of nuclear chromatin

This is usually pale in states of degeneration and/or infection, e.g. herpes infection. The nuclear membrane is blurred. When the cell finally undergoes apoptosis, the nucleus shrinks into a small round hyperchromatic mass.

In reactive processes the nuclear chromatin represents an area of increased biological activity. There is uniform dispersion of small but coarse, similar-sized chromatin granules throughout the nucleus.

The nuclear chromatin of malignant cells is hyperchromatic. This reflects increased DNA and associated proteins. It is the latter that are stained with haematoxylin. Irregularly disposed chromatin clumps in an irregular chromatin background are a reliable diagnostic indicator of malignancy. When the chromatin degenerates, the nuclear outline becomes blurred; however, in well-preserved malignant cells, with increased DNA and associated proteins, hyperchromasia is seen with crisp nuclear outlines. Unequal staining or polychromasia is also a characteristic of rapidly dividing malignant cells which have unequal amounts of DNA and associated proteins.

Nuclear shape

Nuclear pleomorphism is a well recognized feature of malignant cells. The irregularities in the nuclear outline are unpredictable. In malignancy the area of nuclear folding or irregularity faces the cellular membrane away from the centre of the cell, whereas in a benign reactive cell a nuclear indentation always faces the centre of the cell. This discrepancy was considered by the late John K. Frost to be of great discriminating value as a malignant criterion. Indeed, his book on *The Cell in Health and Disease* is an illuminating treatise on all the biological changes seen in cytology.

Multinucleation

Multinucleation can occur in:

- benign cells (osteoclasts)
- reactive processes (multinucleated giant cells, multinucleated reactive epithelial cells)
- malignant cells.

In non-malignant cells the multiple nuclei tend to be uniform in size and shape; however, if they are not, the nuclear outlines are regular and the nuclear chromatin is even. Nucleoli, if they are present, are small, pinpoint and uniform. In malignancy the multinucleation provides an opportunity to compare nuclear features. There is great variation in the nuclear size and shape. The chromatin pattern and staining of these nuclei is uneven.

NUCLEOLI

The presence of nucleoli in benign cells is an indication of protein synthesis, both for intracellular use and for extracellular production. Their number is uniform among a given cell type, and in benign multinucleated cells. In malignancy nucleoli are large and sharply irregular. There is significant variation in the number of nucleoli in each similar malignant cell. It must be stressed that the presence of nucleoli *per se* does not distinguish between cancer and non-cancer. However, there are two exceptions:

- malignant lymphoid cells frequently have large, angular irregular nucleoli; benign lymphoid cells tend to lack them, though immunoblasts can have small uniform nucleoli
- invasive squamous cell carcinoma – with invasion small nucleoli appear as evidence of increased protein production. This may be the first and only cellular change in a dysplastic cell, which is indicative of its potential to invade. The mere presence of nucleoli in these squamous cells should arouse suspicion of possible invasion.

Mitosis

Benign cells and reactive cells may possess normal mitotic figures. Such cells may be quite numerous. The presence of bizarre abnormal mitotic figures, however, is indicative of malignancy. These may be tripolar or quadripolar, or the chromosomes may be dispersed about the cell, creating a 'jellyfish' pattern.

CYTOPLASM

A high N/C ratio is a valuable criterion of malignancy in a large cell. In small lymphoid cells where the nucleus is very large, N/C ratio is not relevant.

For accurate and valid evaluation a cell must have intact cytoplasm. An almost bare nucleus is the remnant of a degenerated cell. A truly bare nucleus with faint wisps of cytoplasm can alert the microscopist to the possibility of abnormality. It should indicate the need for further examination and a second opinion, but it should never be used *per se* to diagnose malignancy. Extremely scant, intact cytoplasm surrounding a large nucleus is a good criterion of malignancy.

In cells which are allowed to spread before fixation, the nucleus stays flattened in the centre and the cytoplasm spreads laterally, most of it lateral to the nucleus (similar to a poached egg). It is often difficult to assess the amount of cytoplasm if the cells have been fixed before spreading, as they round up into a firm ball like a boiled egg. On the other hand, cells which remain embedded in thick mucus cannot spread their cytoplasm, and so it appears scantier than usual. This is why assessment of the quality of fixation *and* background is so important in cytopathology.

The relationship of the nuclear membrane to the cell membrane is often bizarre in cancer cells. The large nucleus adheres closely to the cell membrane in one part of the cell, appearing to have a rim of cytoplasm, while the remainder of the cytoplasm appears 'empty'. Sometimes the cell membrane seems to 'hug' the nuclear membrane of abnormal elongated nuclei.

The late J. K. Frost emphasized the following points at the end of each workshop on *The Cell in Health and Disease*: there is no perfect single criterion of malignancy which, when present, means that cancer is present, or, when it is absent, means that there is no cancer. Neither is there a perfect benign cell. First, determine, by assessing nuclear features, that cancer is present. Second, determine, by assessing the cytoplasmic features, the cell type of cancer present. It is hazardous to reverse this order by using cytoplasmic features as malignant criteria, as no matter how atypical these are they are usually meaningless without assessment of the nuclear criteria. In fact, they can be misleading and invite diagnostic disaster, rather than diagnostic confidence.

REFERENCE

Frost, J.K. *The Cell in Health and Disease: Monographs in Clinical Cytology*, 2nd edn. Berlin, Karger, 1986.

1 Salivary gland tumours

Most neoplastic lesions of the salivary glands are painless and firm (with the exception of adenoid cystic carcinoma, which may be painful owing to perineural involvement). Thus their clinical assessment may be difficult. Other causes of salivary gland enlargement are equally uninformative in their clinical presentation.

Imaging techniques may be helpful in the localization of the mass but not in the specific differentiation of benign and malignant neoplasms. Relatively few large, statistically comparable studies on FNA of salivary gland masses have been published. This may be because histological follow-up is often not available for lesions which are considered benign. These are far more numerous than malignant neoplasms.

Cystic change may be encountered in pleomorphic adenoma (PA), adenoid cystic carcinoma and Warthin's tumour. It is universally accepted that fluid from cystic cavities yields cellular material which exhibits degenerative changes; these are often difficult, if not impossible, to interpret accurately.

There may be marked cellular pleomorphism of stromal cells in pleomorphic adenoma, resulting in false-positive diagnoses. It is helpful to remember that 'commonest things are commoner', e.g. epithelial carcinomas are commoner than sarcomas.

Mucoepidermoid carcinomas are notoriously difficult to assess; however, Cohen *et al.* (1990) recently re-evaluated the cytological criteria for this tumour type and claimed almost 100% accuracy.

It is accepted that the diagnosis of lymphoproliferative lesions in general is not possible on cytology specimens alone; however, as discussed in Chapter 5, it is possible to assess these for the triage management of patients.

PLEOMORPHIC ADENOMA

The unique morphological composition of this tumour usually points to the correct diagnosis. Points to be remembered are:

- the myoepithelial component may be predominant, so that nests of epithelial cells may be difficult to find
- scanning the FNA specimen at very low power yields an overall bird's-eye view of the cellular material and prevents preoccupation with the nuclear pleomorphism seen in the stroma
- when present, tyrosine-like crystalloids are a helpful pointer to the diagnosis, but they are present in very few specimens
- intracytoplasmic mucus may occur. This should not lead to a diagnosis of mucoepidermoid carcinoma if the nuclear features are assessed first, and then attention is paid to the cytoplasmic features.

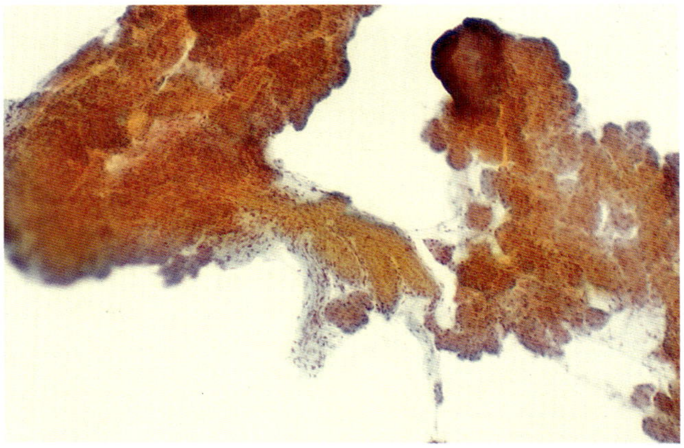

Figure 1.1 FNA of a diffusely enlarged parotid gland in a young adult, showing normal salivary gland acini, sheets of ductal epithelium are present in the centre of the figure and a small amount of loose stromal tissue. Bilateral enlargement of the parotid glands is seen in bulimia and alcoholism; no structural or histopathological changes of the functional glandular cells are seen.

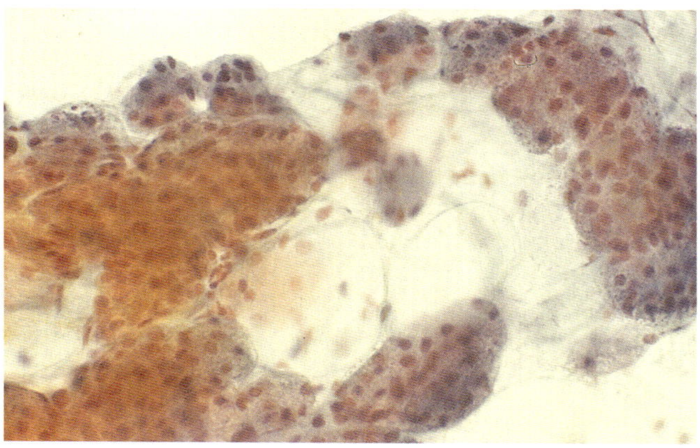

Figure 1.2 'Fatty infiltration of the parotid'. Normal salivary gland acini are shown, separated by benign adipose tissue. This phenomenon is often seen in FNA specimens from older patients who are noted to have asymptomatic enlargement of the parotid, or displacement of the parotid gland inferiorly into the submandibular area. It is also seen in patients who are malnourished for any reason, e.g. anorexia, cachexia, alcoholism. Note the uniform shape of the cells, which are arranged in acini. The cytoplasm is plentiful and contains finely granular material. The dark central nuclei are uniform in size and shape.

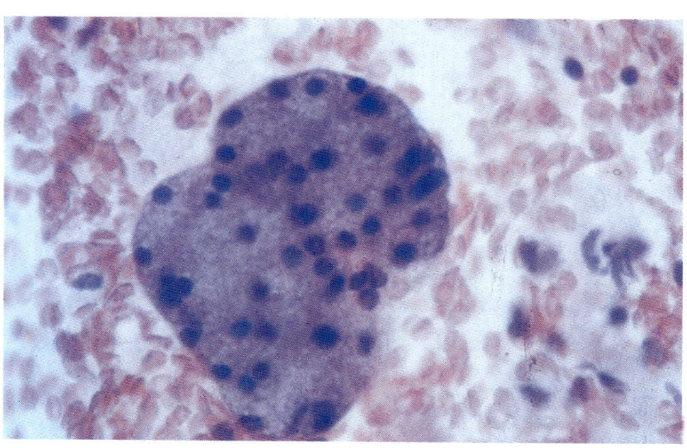

Figure 1.3 High-power view of normal salivary gland acini. Note the peripheral disposition of the nuclei, the plentiful cytoplasm and the cohesive cells forming each acinus. Cellular outlines are clearly defined. ×400.

SALIVARY GLAND WITH LYMPHOID INFILTRATION AND GRANULOMA FORMATION

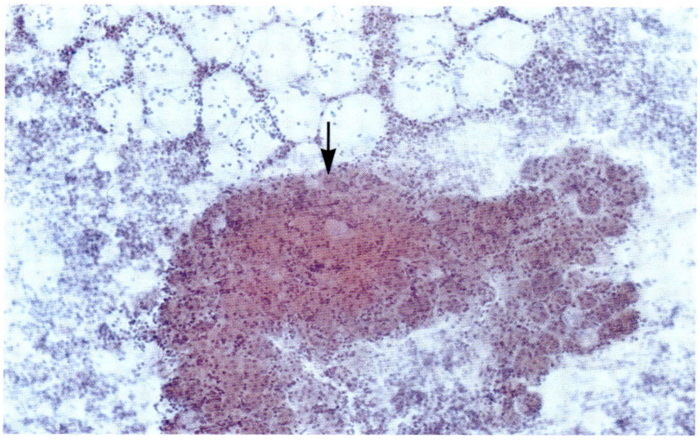

(a)

Figure 1.4 (a) Lymphoid infiltration of salivary gland tissue. A microbiopsy of salivary gland tissue is present adjacent to adipose tissue. There is prominent infiltration by loose-lying lymphoid cells, with apparent destruction of the acinar pattern (arrow). ×100.

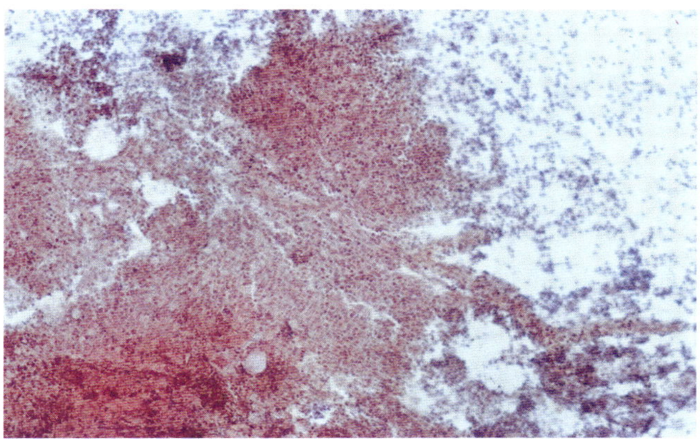

(b)

Figure 1.4 (b) This FNA of a salivary gland shows marked lymphoid infiltration and granuloma formation. An acinar pattern is not seen. ×100.

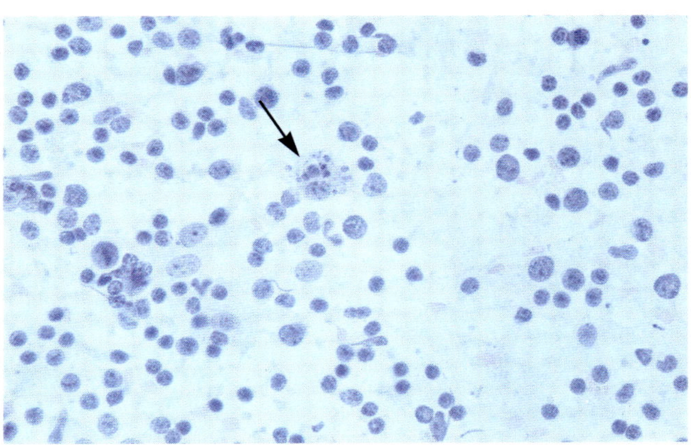

(c)

Figure 1.4 (c) High-power view of lymphoid cells from the salivary gland infiltrate seen in (a) and (b). Note that the lymphoid cells show varying degrees of maturity, indicating a reactive process. A 'tingible-body' macrophage is present (arrow). ×400.

PLEOMORPHIC ADENOMA

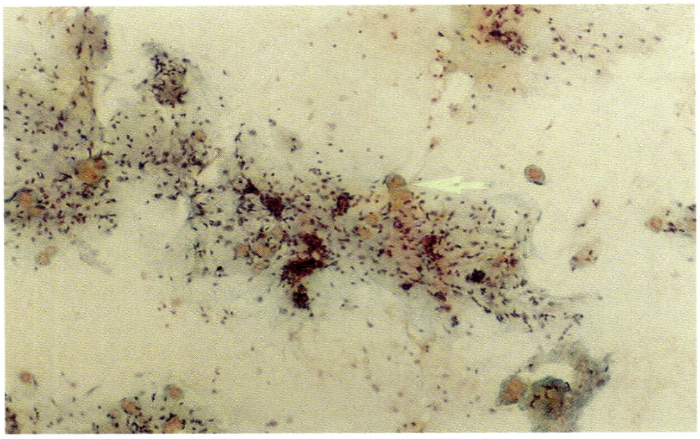

Figure 1.5 FNA of a parotid mass showing groups of tightly clustered epithelial cells lying in a loose, fairly cellular stroma. Scattered orangeophilic refractile crystals are present; these are called tyrosine crystalloids, though in reality they are composed of a mixture of tyrosine and other substances. They are regarded as being pathognomonic of pleomorphic adenoma. The crystals are usually dissolved out during histological processing and are therefore rarely seen in H&E sections. ×100.

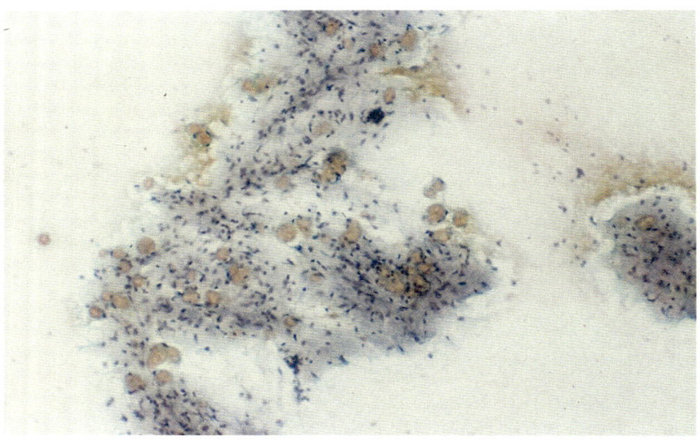

Figure 1.6 Low-power view of loose stromal tissue, with tyrosine-like crystalloids, from a pleomorphic adenoma. ×100.

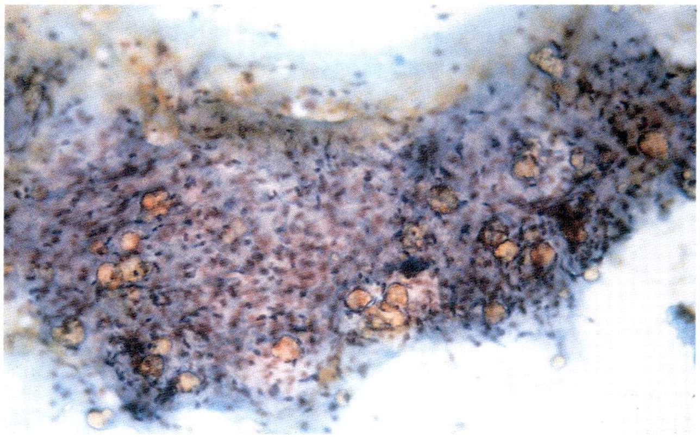

Figure 1.7 FNA of this parotid mass shows very cellular stromal tissue with scattered, brightly orangeophilic tyrosine-like crystalloids. No epithelial cells are present in this field. The specimen was from a pleomorphic adenoma with a predominant myoepithelial component. Tyrosine crystalloids are seen in less than 10% of FNA specimens of pleomorphic adenoma. ×400.

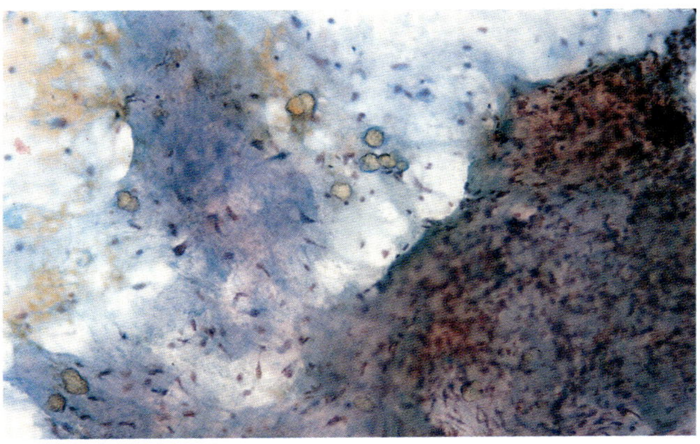

Figure 1.8 Microbiopsies of loose metachromatic stroma adjacent to very cellular stroma and a small cluster of epithelial cells which is almost obscured by the stromal tissue. The stromal cells may be stellate and show quite a marked degree of pleomorphism. A few loose-lying crystalloids are seen. ×400.

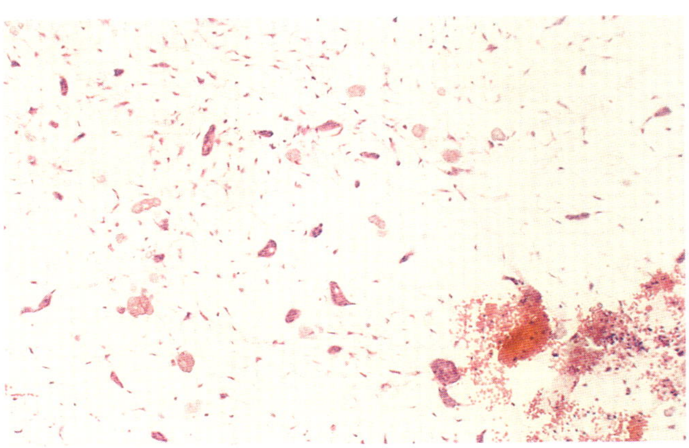

Figure 1.9 Histological section of a pleomorphic adenoma showing the nests of epithelial cells in a loose myxoid stroma. Tyrosine crystalloids are seen in this section as pale pink, amorphous crystalline material. H&E ×100.

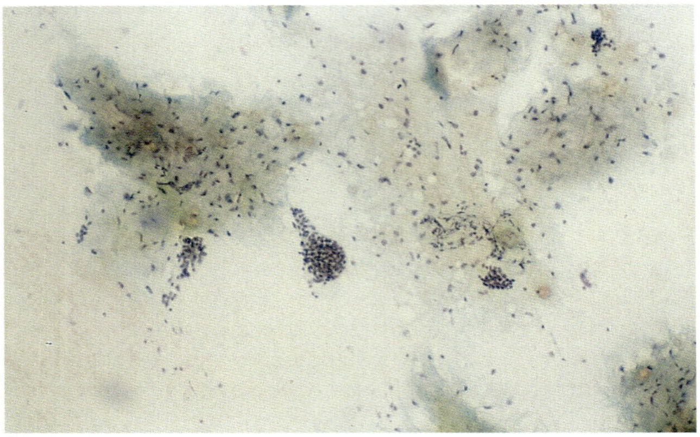

Figure 1.10 FNA of a pleomorphic adenoma: clusters of epithelial cells resembling a swarm of bees are lying in loose metachromatic stroma. Pale tyrosine crystalloids are also present. ×100.

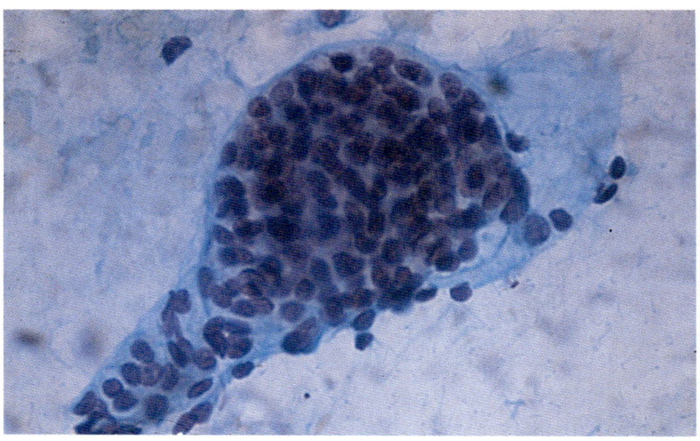

Figure 1.11 High-power view of the cluster of epithelial cells seen in the low-power view in Figure 1.10. The 'swarm of bees' appearance is quite characteristic of pleomorphic adenoma. The nuclei are very uniform in size and shape. ×400.

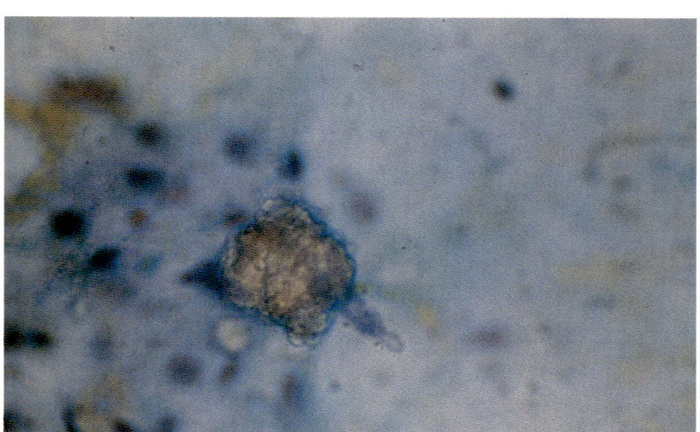

Figure 1.12 High-power view of refractile tyrosine-like crystalloid resembling the petals of a flower. ×400.

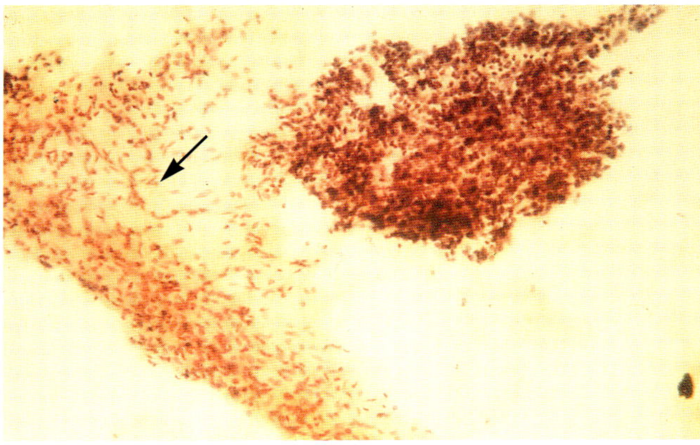

Figure 1.13 Microbiopsies of tissue from a pleomorphic adenoma. Loose stromal tissue and sheets of three-dimensional epithelial cells are shown. No crystals were found in this aspirate. Note the leashes of capillaries in the stromal tissue (arrow). ×100.

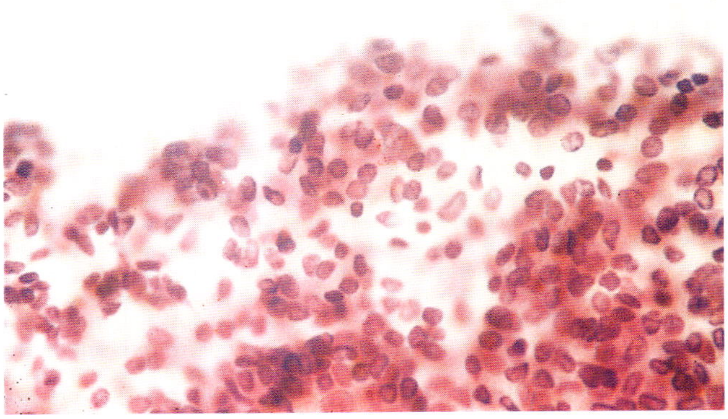

Figure 1.14 High-power view of epithelial cells. Note the fairly uniform size of the cells, even though they have a loose arrangement. ×400.

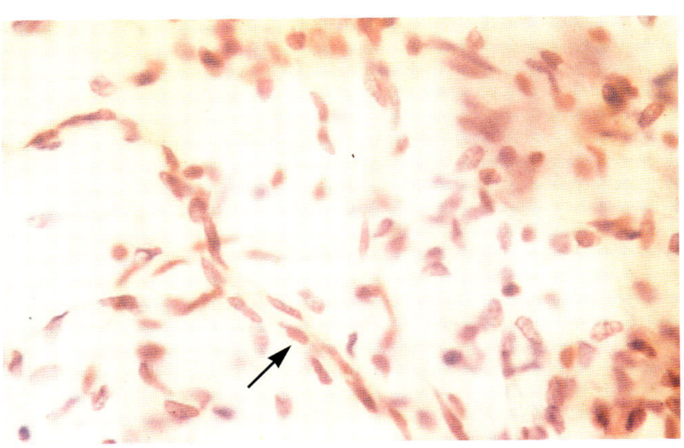

Figure 1.15 High-power view of stromal tissue from a pleomorphic adenoma, showing the spindle-shaped nature of the nuclei lying in a translucent stroma. Note, too, the capillary vessel (arrow). ×400.

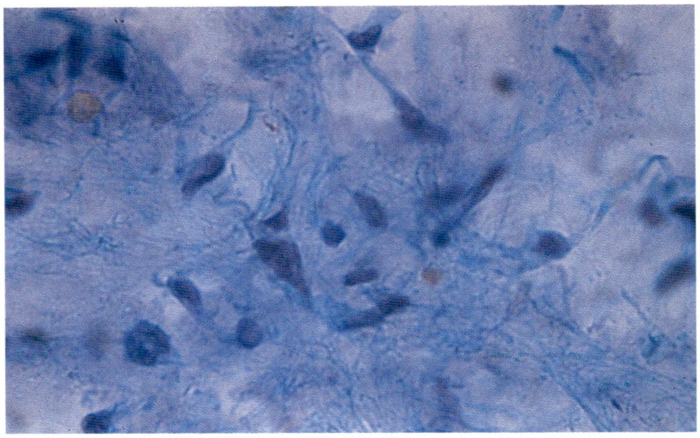

Figure 1.16 High-power view of stromal cells. Many have stellate nuclei and fibrillary cytoplasm with apparent cytoplasmic tails. ×400.

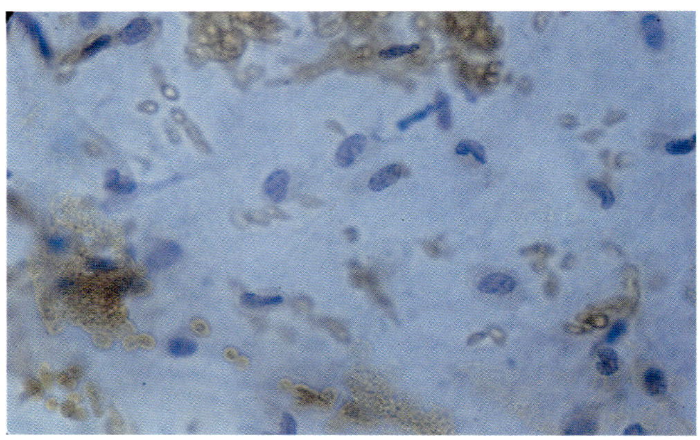

Figure 1.17 Another high-power view of markedly atypical stromal cells from a benign pleomorphic adenoma. ×400.

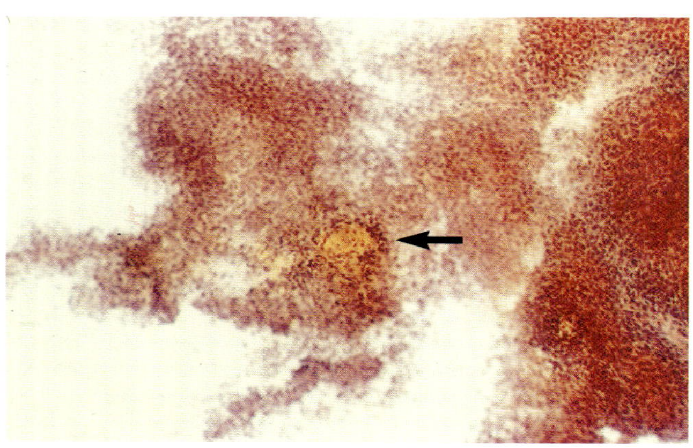

Figure 1.18 Low-power view of a pleomorphic adenoma with predominant myoepitheliomatous component. Note the squamous pearl (arrow). These may be associated with squamous metaplasia occurring in the epithelial duct-lining cells. ×100.

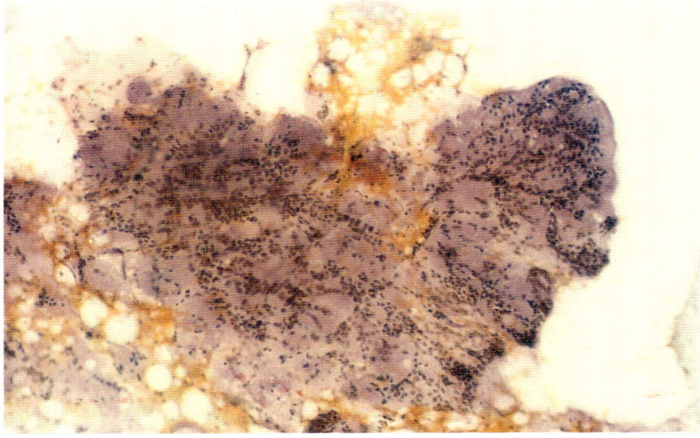

Figure 1.19 A lobulated microbiopsy of metachromatic tissue from a pleomorphic adenoma with a predominant myoepithelial component. ×100.

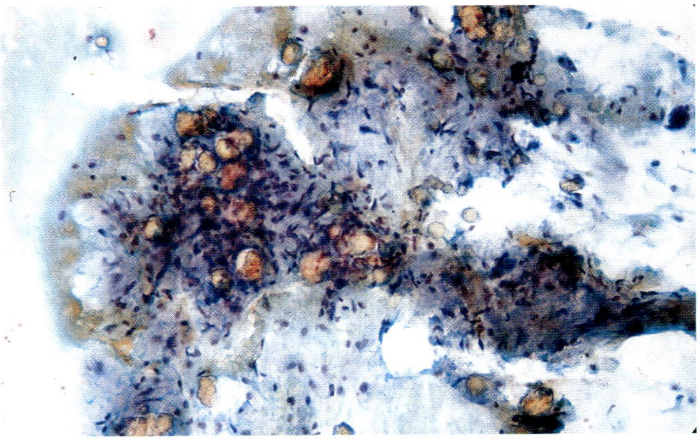

Figure 1.20 Numerous tyrosine-like crystalloids lying in metachromatic cellular stroma. ×100.

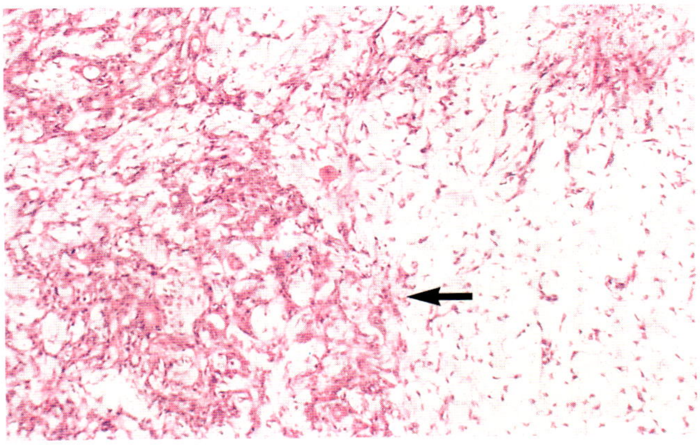

Figure 1.21 Histological section of a pleomorphic adenoma with a predominant myoepitheliomatous component. No crystalloids are seen. H&E ×100.

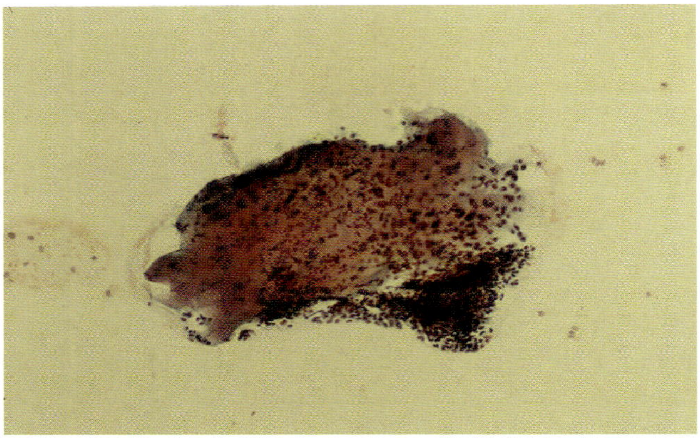

Figure 1.22 This FNA of a submandibular mass yielded only a single microbiopsy of tissue, composed of small, crowded epithelial cells and metachromatic stromal tissue. The possibility of a pleomorphic adenoma, rather than a metastatic carcinoma, was suggested. ×100.

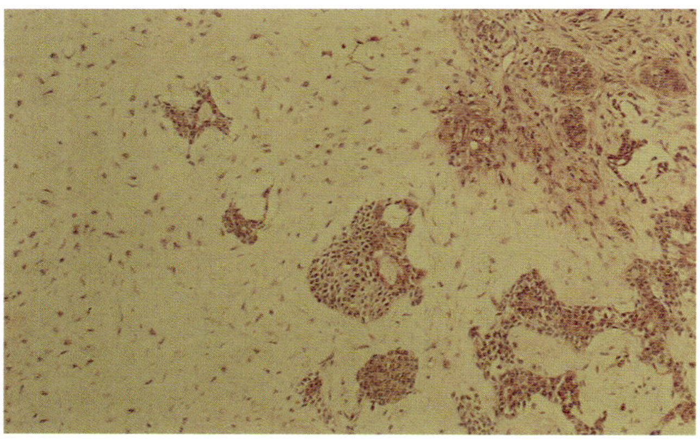

Figure 1.23 The mass shown in Figure 1.22 was excised; histology revealed a benign pleomorphic adenoma with prominent myoepithelial component. H&E ×100.

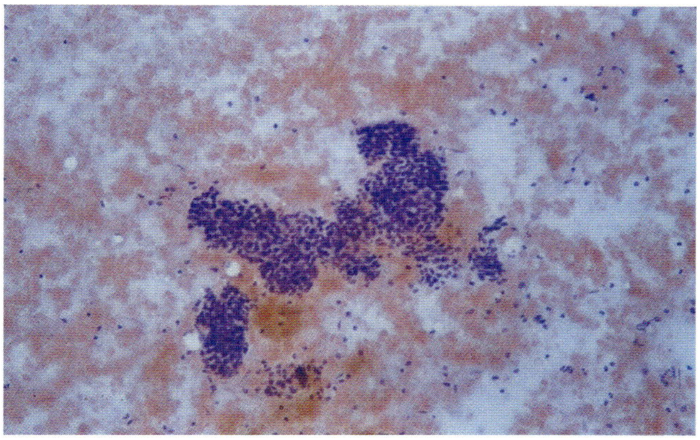

Figure 1.24 FNA of a submandibular mass which yielded predominantly epithelial tissue consisting of clusters of tightly cohesive, hyperchromatic epithelial cells in a bloodstained background. No stromal tissue was present. ×100.

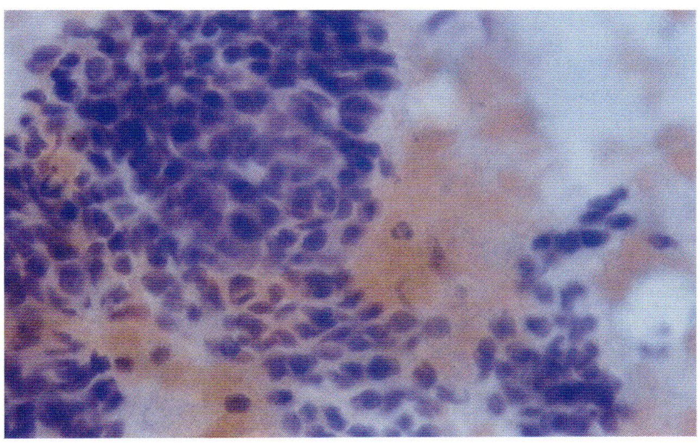

Figure 1.25 High-power view of the epithelial cells shown in Figure 1.24, which emphasizes their tight cohesiveness or 'swarm of bees' appearance. ×100.

WARTHIN'S TUMOUR (ADENOLYMPHOMA)

FNA of adenolymphoma often yields very scanty material owing to the cavitary nature of the lesion. Often, only very scanty isolated groups of epithelial cells may be found lying in a background of mucoid material containing lymphocytes. For this reason it is difficult to find a suitable representative field for photography.

Figure 1.26 FNA from a Warthin's tumour. A small isolated sheet of epithelial cells is seen lying in a background of lymphoid cells and thick mucoid material. ×40.

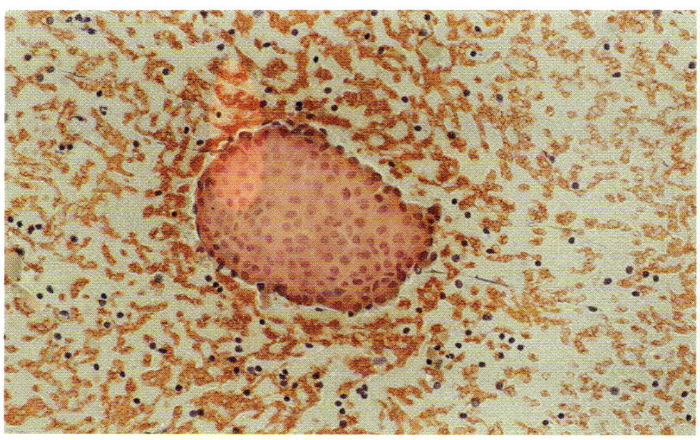

Figure 1.27 High-power view of a sheet of epithelial cells with a 'honeycomb' arrangement, from a Warthin's tumour. ×400.

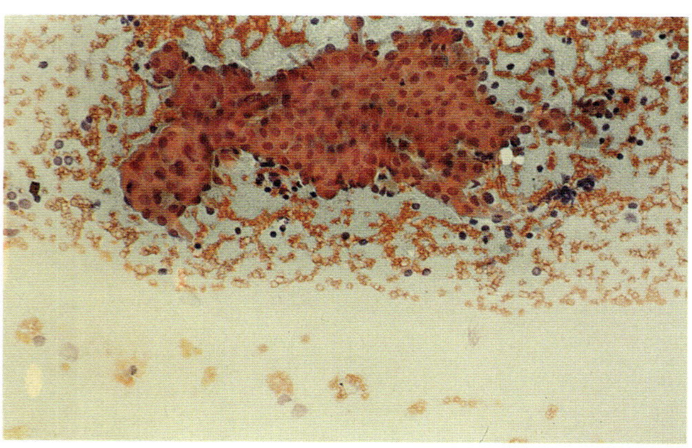

Figure 1.28 Isolated sheet of fairly uniform cells, with a 'honeycomb' arrangement, lying in a background of scattered lymphocytes and erythrocytes. ×400.

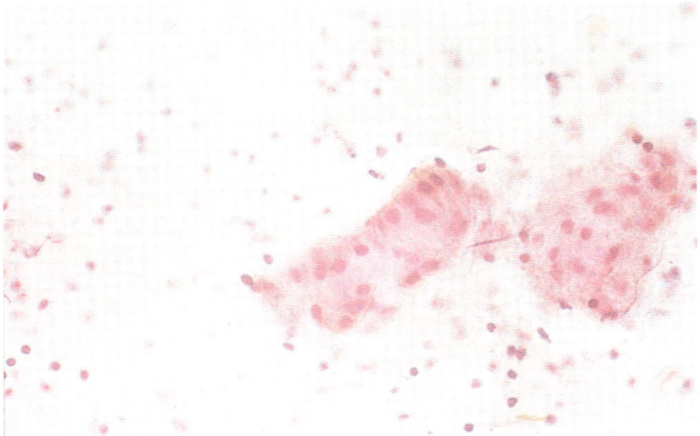

Figure 1.29 High-power view of the sheets of epithelial cells from an adenolymphoma, showing the columnar nature cf the cells. Note the even cellular border. Lymphocytes are present in the mucoid background. ×400.

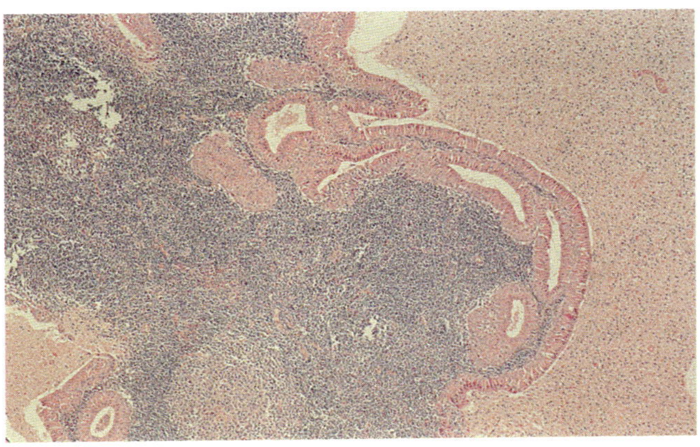

Figure 1.30 Histological section of the mass shown in Figure 1.29 confirmed a Warthin's tumour composed of undulating trabeculae of tall eosinophilic columnar cells, which are intimately associated with lymphoid tissue. Note the amorphous cellular debris in the cystic spaces. H&E ×40.

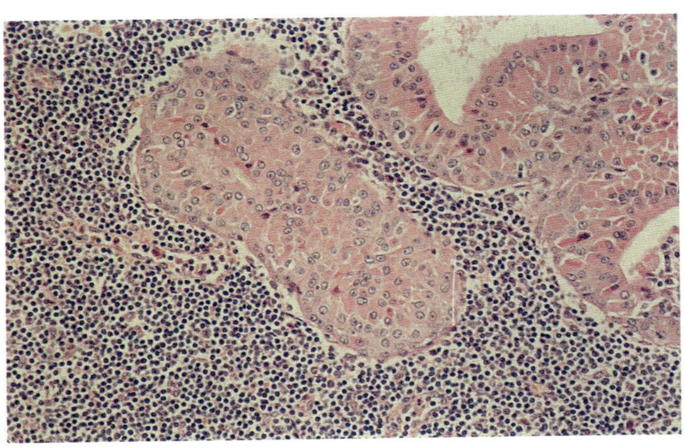

Figure 1.31 High-power view of the histological section, emphasizing the tall uniform nature of the eosinophilic epithelial cells and their close association with lymphocytes. H&E ×100.

ADENOID CYSTIC CARCINOMA

The proclivity of this tumour to infiltrate and spread along the perineural sheath is probably responsible for the fact that the patient often presents with a painful, slow-growing mass in the region of a salivary gland. This tit-bit of clinical information may lead the pathologist to the correct diagnosis in difficult cases. The cribriform arrangement of cells with fairly uniform, peripherally placed nuclei around central globules of hyaline, is often a striking feature on FNA specimens; however, in some cases this pattern may not be so obvious as the cells seem to be lying in sheets, not associated with hyaline material. In less well differentiated cases the differentiation of adenoid cystic carcinoma from adenocarcinoma may not be possible. The resemblance to mucoepidermoid carcinoma may also be confusing. In these cases it is not usually the responsibility of the cytopathologist to classify the neoplasm, but it is satisfying to be specific and correct, occasionally.

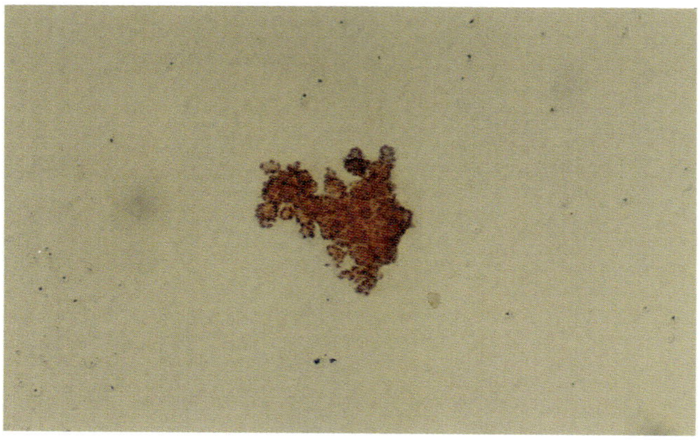

Figure 1.32 FNA of an adenoid cystic carcinoma: low-power view of a microbiopsy lying in a clear background, showing an obvious cribriform pattern. ×100.

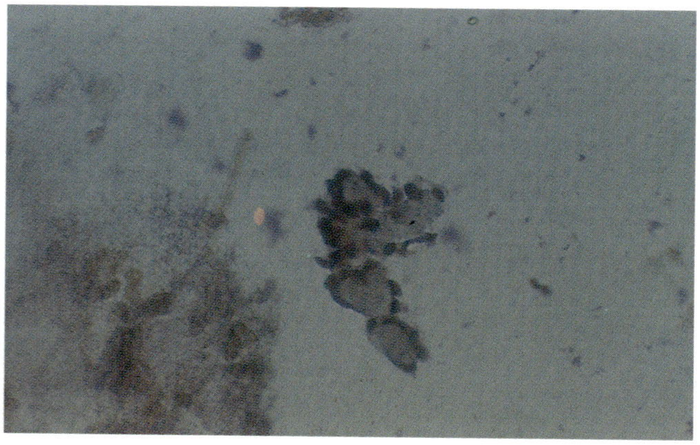

Figure 1.33 FNA of an adenoid cystic carcinoma. The uniform peripheral nuclei are arranged around globules of amorphous material which probably corresponds to the hyaline material seen on histological sections. Note the clear background. ×400.

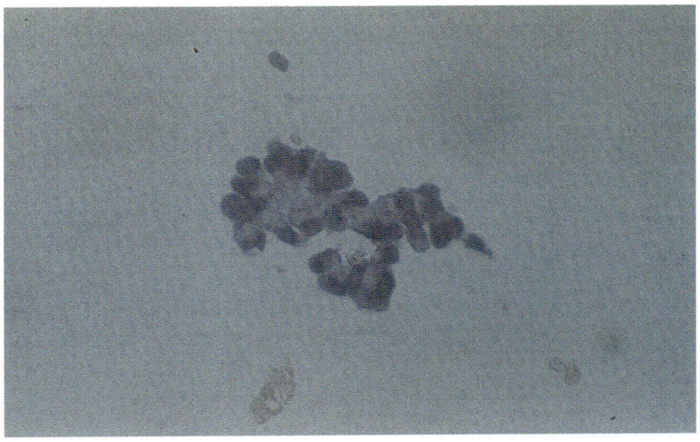

Figure 1.34 Adenoid cystic carcinoma. The nuclei in this group show moderate pleomorphism; however, they are still in a peripheral position. ×400.

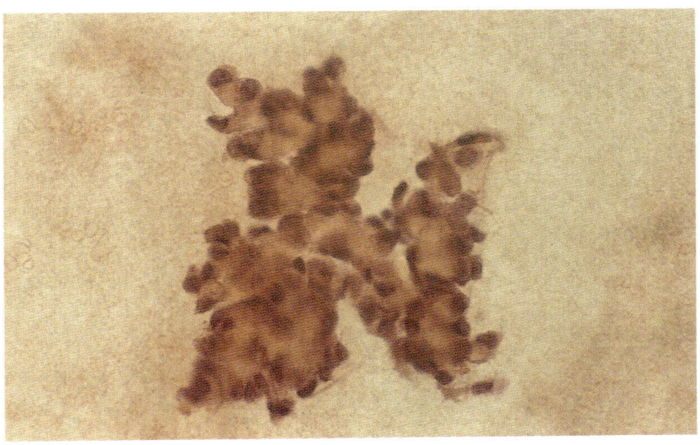

Figure 1.35 Adenoid cystic carcinoma: high-power view. The cribriform arrangement of these cells is very marked. ×400.

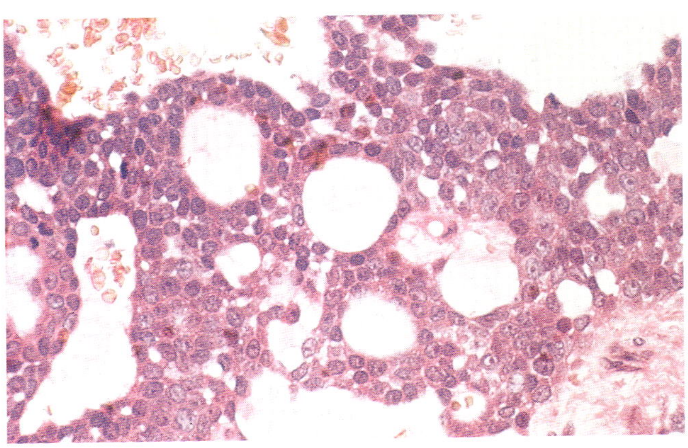

Figure 1.36 Histological section showing a typical adenoid cystic carcinoma. H&E ×400.

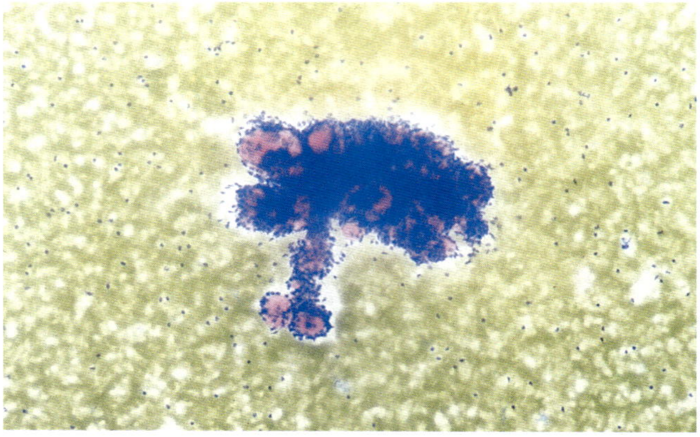

Figure 1.37 This is an air-dried specimen of an FNA of an adenoid cystic carcinoma stained with Giemsa stain. It demonstrates beautifully the adenoid cystic nature of the tumour. The central carmine mass of hyaline contrasts with the intensely basophilic peripheral nuclei. ×100.

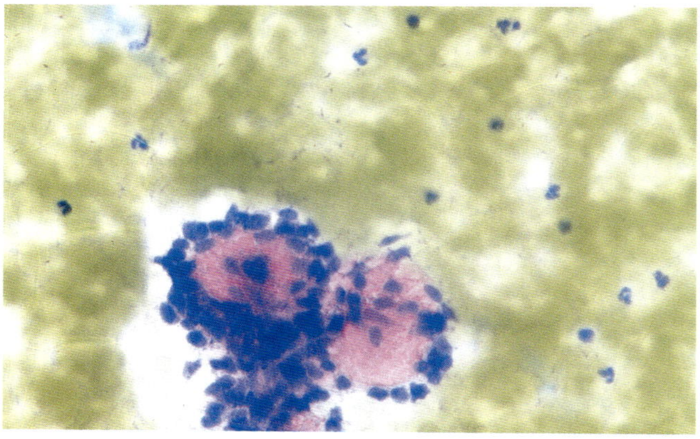

Figure 1.38 A high-power view of the Giemsa-stained preparation shown in Figure 1.37. ×400.

MUCOEPIDERMOID CARCINOMA

Possible pitfalls in the diagnosis of this neoplasm on FNA specimens are discussed above (see also page 121). It is not a common tumour. It is generally found in elderly patients, though the author has seen an FNA specimen from a mucoepidermoid carcinoma in a 5-year-old who had received therapy for medulloblastoma.

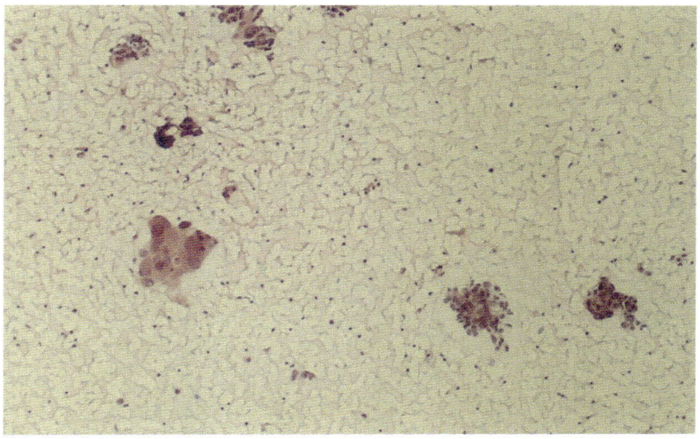

Figure 1.39 FNA of a mucoepidermoid carcinoma. Scanty groups of fairly cohesive cells, some of which are pleomorphic, are seen in a 'mucoid' background. This has a unique pattern which has been called 'tigroid' substance because of its resemblance to the stripes on a tiger. There are no features of either pleomorphic adenoma or adenoid cystic carcinoma. The background suggests that this neoplasm could be a mucoepidermoid carcinoma. ×400.

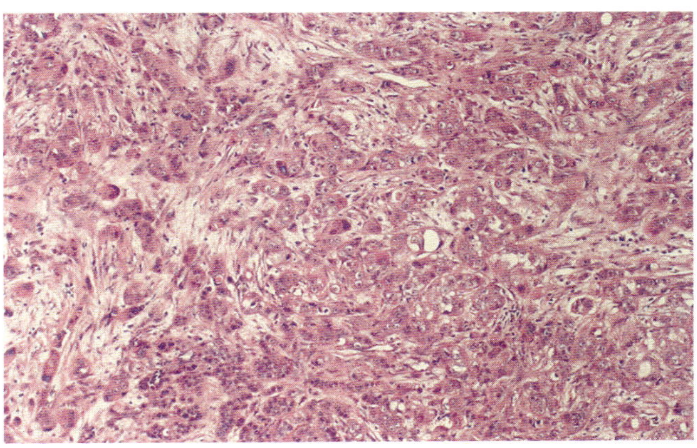

Figure 1.40 Histological section of the excised mass shows a poorly differentiated mucoepidermoid carcinoma. ×100.

REFERENCES AND FURTHER READING

Atula, T., Grénman, R., Laippala, P. and Klemi, P.-J. (1995) Fine-needle aspiration cytology of submandibular gland lesions. *J. Laryngol. Otol.* **109**: 853–858.

Bottles, K., Ferrell, L.D. and Muller, T.R. (1984) Tyrosine crystals in FNA of pleomorphic adenoma of the parotid gland. *Acta Cytol.* **28**: 490–496.

Chan, M.K.M., McGuire, L.J., King, W. and Li, J.C.K. (1992) Cytodiagnosis of 112 salivary gland lesions. Correlations with histologic and frozen section diagnosis. *Acta Cytol.* **36**: 353–363.

Cohen, M.B., Fisher, P.E., Holly, E.A., Ljung, B.-M., Löwhagen, T. and Bottles, K. (1990)Fine-needle aspiration biopsy diagnosis of mucoepidermoid carcinoma. *Acta Cytol.* **34**: 43–49.

Klijanienko, J. and Viehl, P. (1996) Fine-needle sampling of salivary gland lesions. I Cytology and histology correlations of 412 cases of pleomorphic adenoma. *Diagn. Cytopathol.* **14**: 195–200.

Kocjan, G., Nayagam, M. and Harris, M. (1990) Fine-needle aspiration cytology of salivary gland lesions: advantages and pitfalls. *Cytopathology* **1**: 269–275.

Layfield, L.J. and Glasgow, B.J. (1991) Diagnosis of salivary gland tumours by fine-needle aspiration cytology: a review of clinical utility and pitfalls. *Diagn. Cytopathol.* **7**: 267–272.

Learmonth, G.M., Sellars, S. and Duff, J. (1990) Cytopathology of the head and neck as an aid to the management of ENT patients. (Abstract.) *Proc. Swed. Otolaryngol. Soc.* **2**: 143–144.

Orell, S.R. and Nettle, W.J.S. (1998) Fine-needle aspiration biopsy of salivary gland tumours: problems and pitfalls. *Pathology* **20**: 332–337.

Palma di, S. (1996) FNA of myoepithelioma of salivary gland. *Cytopathology* **7**: 357–365.

Roland, N.J., Caslin, A.W., Smith, P.A., Turnbull, L.S., Panarese, A. and Jones, A.S. (1993) Fine-needle aspiration cytology of salivary gland lesions reported immediately in a head and neck clinic. *J. Laryngol. Otol.* **107**: 1025–1028.

Weinberger, M.S., Rosenberg, W.W., Meurer, W.T. and Robbins, K.T. (1992) Fine-needle aspiration of parotid gland lesions. *Head Neck* **14**: 483–487.

Young, J.A., Smallman, L.A., Thompson, H., Proops, D.W. and Johnson, A.P. (1990) Fine-needle aspiration cytology of salivary gland lesions. *Cytopathology* **1**: 25–33.

2 Lymph nodes – metastatic lesions

The commonest lesions found on FNA samples from lymph nodes in the neck are metastatic carcinomas, usually from the head and neck region, lung or breast. However, metastases can present from distant sites such as the cervix, prostate and rectum; indeed, they may be the first clinical presentation of these primary lesions. Metastatic malignant melanoma may pose a difficult problem, as the melanoma cells may contain very little or no pigment and a primary lesion may not be obvious, e.g. occult pigmented lesions on the scalp obscured by hair. The differential diagnosis of metastatic malignancy in neck nodes requires careful consideration of the clinical background (Theunnissen *et al.*, 1993). Sometimes the primary lesion is known, but often it is not, as the enlarged neck node may be the initial presenting complaint. The duty of the cytopathologist is to guide the clinician to the quickest, cheapest and easiest way to a final diagnosis or to definitive management of the patient. Suggestions concerning blood tests, X-rays and other clinical tests may be made at the outpatient clinic, thereby speeding up further investigations and management.

In many instances the cytopathologist can offer confirmation of the diagnosis in cases with known primaries, and indeed may point to the presence of another pathological process in the absence of a metastasis (Schwarz *et al.*, 1990). This is very useful in the triage management of outpatients. It must be remembered that patients with a known primary malignancy can also have some other disease process, e.g. another neoplastic process such as lymphoma, or an inflammatory process such as tuberculosis, toxoplasmosis or cat-scratch disease. Aspiration of necrotic material from a neck node may indicate a necrotic centre in a metastatic keratinizing squamous carcinoma, an infected cyst, an abscess, or florid acute tuberculosis.

When apparent pus or necrotic material is aspirated (Strickler *et al.*, 1987), a slide can be put aside for Gram or Ziehl–Neelsen staining (depending on the clinical situation), and a small amount of material may be sent for culture. The request for bacteriological culture may be subsequently cancelled if it is found not to be necessary.

The author has never made an attempt definitively to classify lymphoproliferative disorders on FNA material, although suggestions as to the nature of the process were made. In clinical practice, it sufficed to suggest the possibility of a lymphoproliferative disorder and to direct the patient for further investigation at the lymphoma clinic (see Chapter 5).

METASTATIC SQUAMOUS CARCINOMA

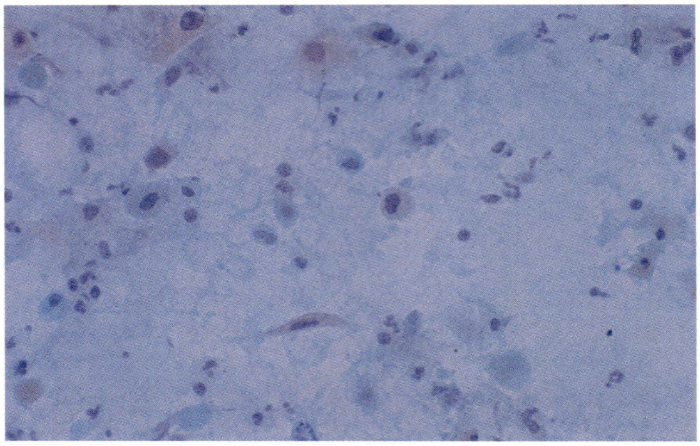

Figure 2.1 FNA of metastatic squamous carcinoma in a neck node from a female patient. Note the loose-lying, round- and spindle-shaped cells with long cytoplasmic tails, which give a clue to the diagnosis of metastatic squamous carcinoma. The cellular material is pale-staining owing to degenerative changes; these are often seen in FNA specimens from metastatic nodes harbouring squamous carcinoma. The primary lesion was found in the cervix. ×100.

This patient had initially presented with weight loss and was found to have enlarged lymph nodes. Further investigation was suggested, starting with the least invasive, least expensive test, a cervical smear. Evidence of cervical carcinoma was found.

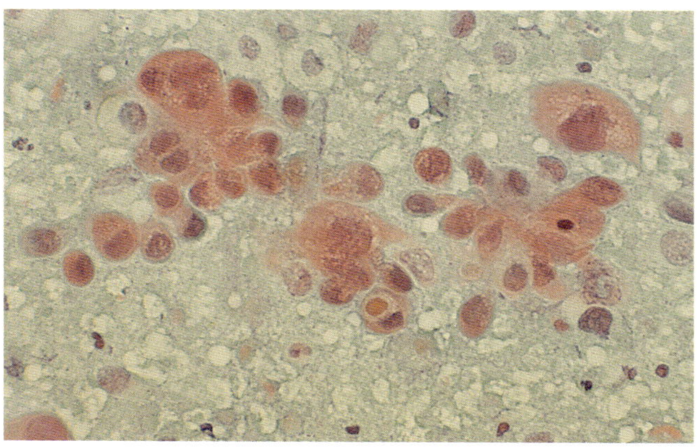

Figure 2.2 FNA of a neck node. High-power view of metastatic keratinizing squamous carcinoma showing degenerative changes in the pleomorphic malignant cells. The cytoplasm is eosinophilic and pleomorphic. Many small intracellular vacuoles are present, indicating degeneration. Intracytoplasmic keratin is noted in one cell (dyskeratosis). ×400.

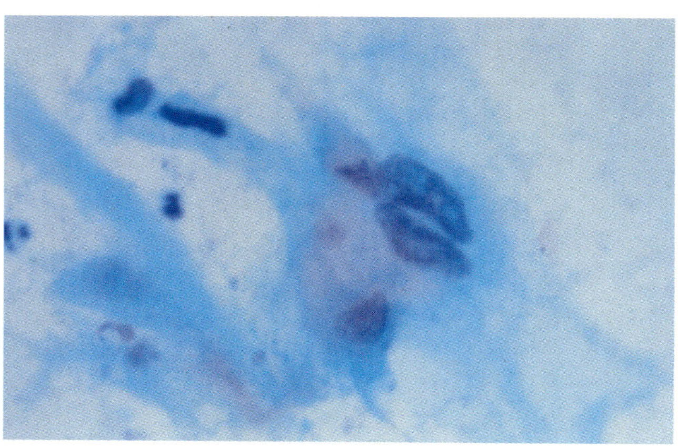

Figure 2.3 FNA of a neck node. Metastatic squamous carcinoma from the oesophagus. The cells show a marked degree of pleomorphism, with variations in shape that are characteristic of squamous neoplasia. Note the elongated spindle cell ('fibre cell'). ×400.

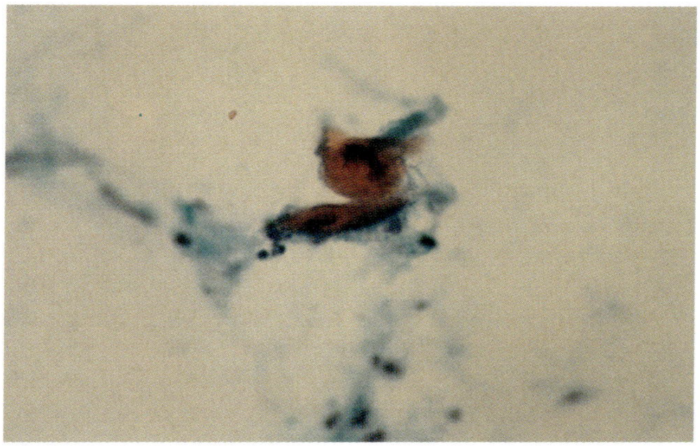

Figure 2.4 Intensely orangeophilic, keratinized squamous cells from a neck node in a patient with known keratinizing squamous carcinoma of the lung. ×400.

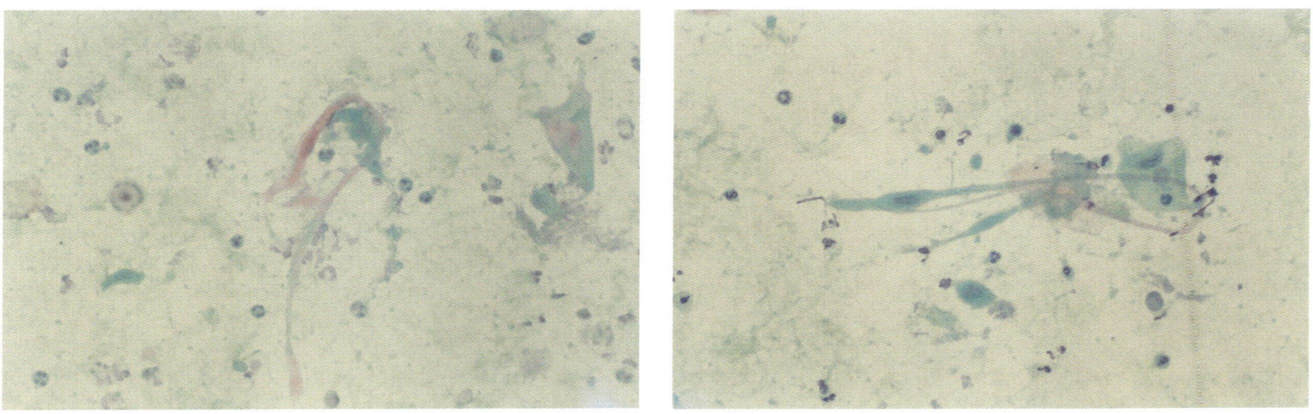

(a) (b)

Figure 2.5 (a) and (b) Extremely long cytoplasmic tails in keratinized (eosinophilic) cells and non-keratinized cells (basophilic). FNA of neck node with metastatic squamous carcinoma from the larynx. ×400.

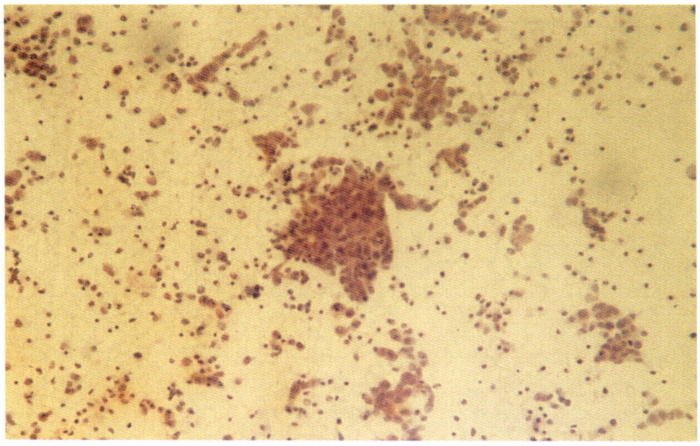

Figure 2.6 Metastatic poorly differentiated carcinoma: a very cellular aspirate composed of clusters of small cells with scanty eosinophilic cytoplasm. Similar loose-lying cells are present in the background. The cells are not intensely hyperchromatic and chromatin smearing is not a feature, suggesting that this is not a small-cell (oat-cell) carcinoma. The patient had a large neoplastic lesion in the lung, with skeletal metastases. ×100.

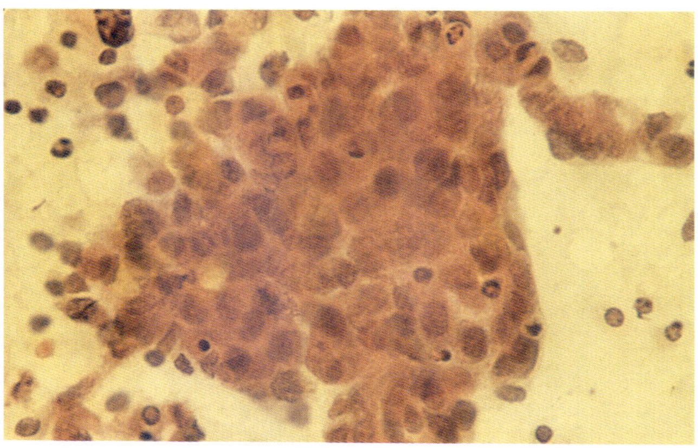

Figure 2.7 High-power view of the specimen shown in Figure 2.6 to show the cellular detail – the nuclei are quite pleomorphic. Occasional cells show apoptosis. Mitoses are present. There are no features to suggest whether this is metastatic adenocarcinoma or squamous carcinoma. ×400.

METASTATIC SMALL-CELL (OAT-CELL) CARCINOMA

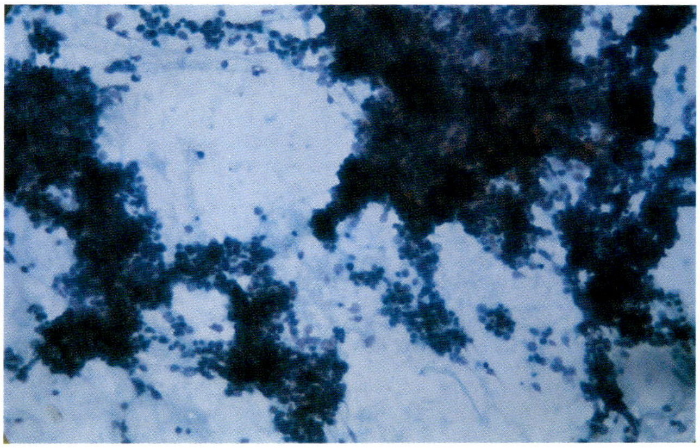

Figure 2.8 Metastatic small-cell (oat-cell) carcinoma from the lung. Low-power view of an FNA showing sheets and microbiopsies of small, crowded, intensely hyperchromatic/basophilic cells which are characteristic of 'oat cells' (small cells). ×100.

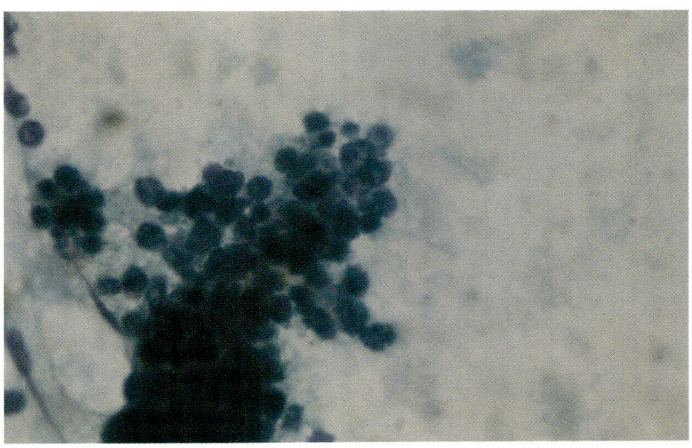

Figure 2.9 Cluster of small, crowded, intensely hyperchromatic cells, with blue-black staining. These are often referred to as 'ink-blot' cells, and are diagnostic of small-cell carcinoma. ×400.

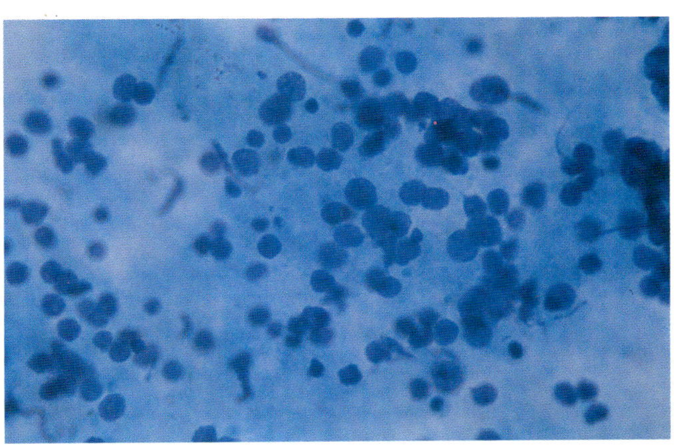

Figure 2.10 Another view of metastatic oat cells, emphasizing their 'ink-blot' appearance. Small groups of cells show moulding which is a feature of small-cell carcinoma (neoplastic lymphoid cells are usually single lying).

METASTATIC CARCINOMA FROM THE BREAST

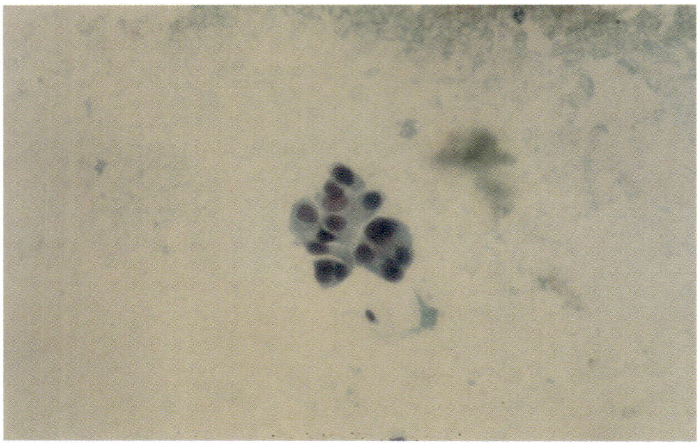

Figure 2.11 Metastatic carcinoma from the breast. A discrete cluster of cells with plentiful cytoplasm and eccentric nuclei lying in a clear background. Although these appearances suggest a lobular carcinoma, it is not possible to categorize the different types of breast carcinoma accurately on cytology. ×400.

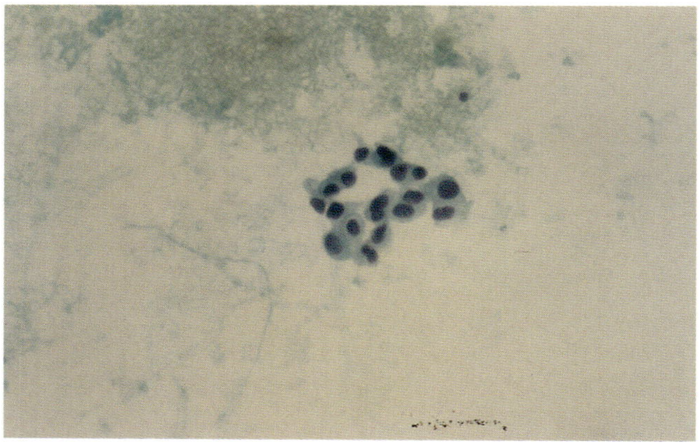

Figure 2.12 Another group of cells with eccentric nuclei and a linear, circular arrangement. This appearance is often seen in lobular carcinoma of the breast. ×400.

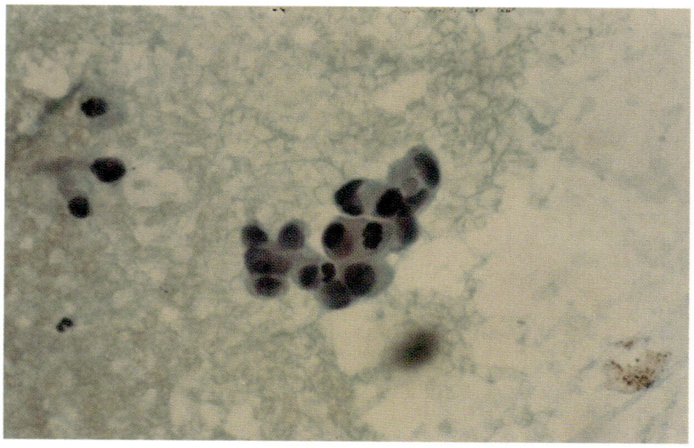

Figure 2.13 A three-dimensional cluster of cells with plentiful cytoplasm and eccentric nuclei. A mitotic figure is present. ×400.

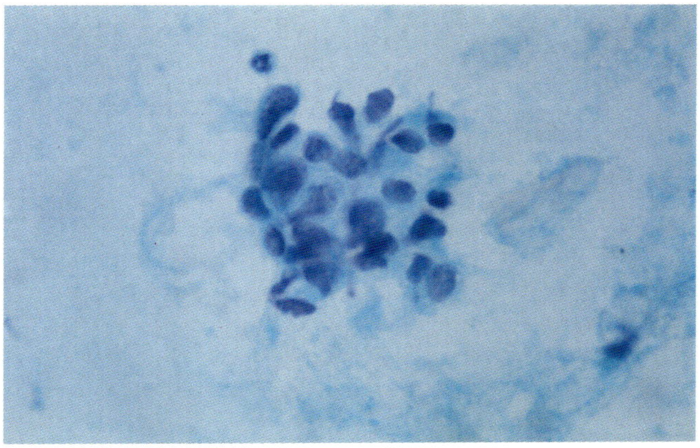

Figure 2.14 Metastatic ductal carcinoma from the breast: a group of pleomorphic cells with plentiful cytoplasm, confirming the presence of metastatic spread from a primary breast carcinoma. ×400.

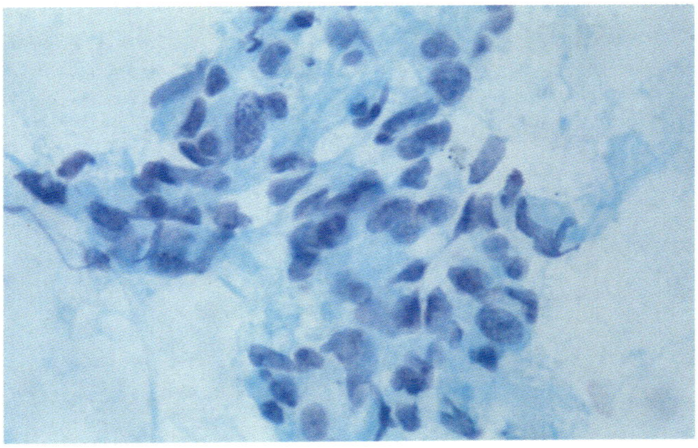

Figure 2.15 High-power view of metastatic duct carcinoma showing disarray of cells and marked pleomorphism of nuclei, with variations in size and shape and in the number of nucleoli. ×400.

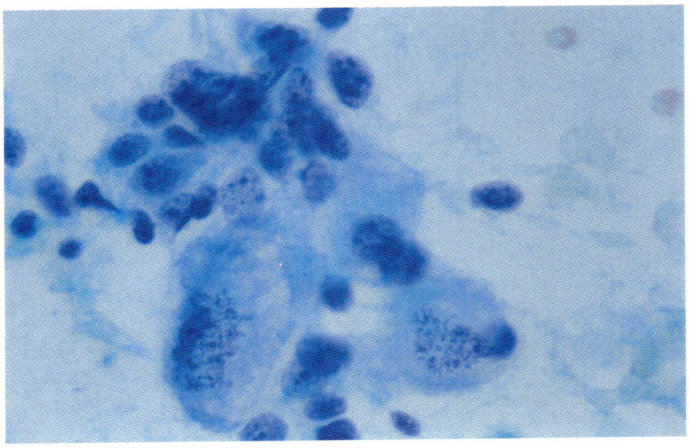

Figure 2.16 Metastatic duct carcinoma in a patient who had had radiotherapy. Very large, bizarre pleomorphic cells with plentiful 'feathery' cytoplasm and nuclei with markedly disturbed chromatin are shown. Multiple prominent nucleoli are present, as well as abnormal mitotic figures. Individual chromosomes can be discerned in one cell. It is unknown whether these appearances are due to radiotherapy or not. ×400.

METASTASES FROM OTHER SITES

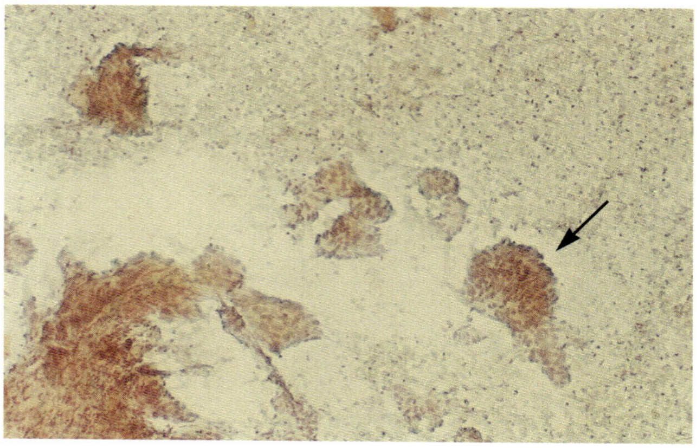

Figure 2.17 Metastatic adenocarcinoma from rectum. FNA of a neck node from a 29-year-old male with a history of familial polyposis of the colon. Low-power view showing a very cellular aspirate with three-dimensional clusters of cells compatible with metastatic adenocarcinoma. Note that many of the groups show cellular and nuclear uniformity amounting to a 'honeycomb' pattern, whereas others show more pleomorphism. Necrotic material is present in the background. These features are often seen in well-differentiated intestinal adenocarcinoma. ×100.

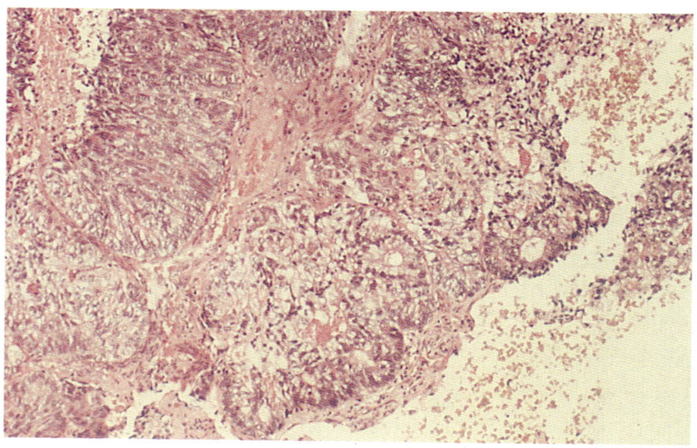

Figure 2.18 Trucut biopsy of the neck node confirmed the presence of a metastatic colonic type of adenocarcinoma. H&E ×100.

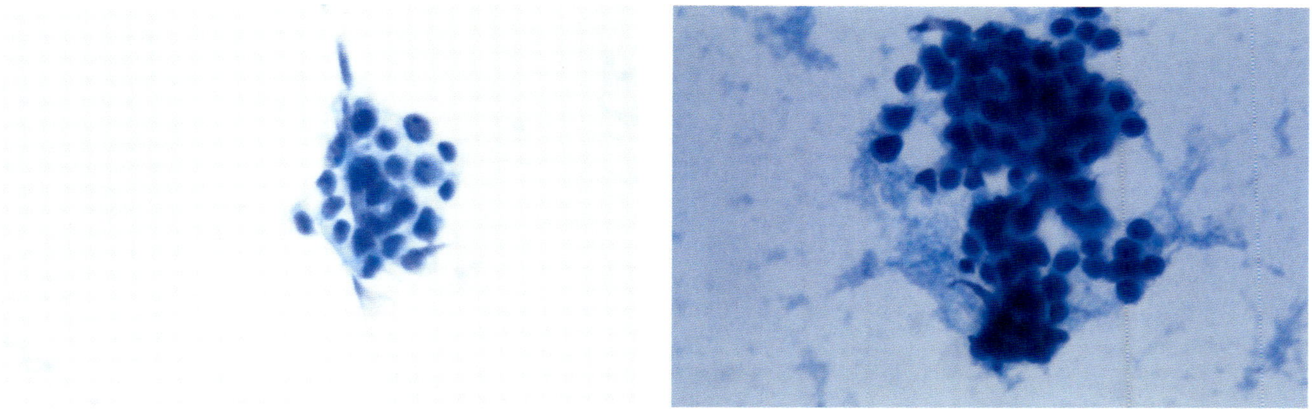

(a) (b)

Figure 2.19 (a) and (b) FNA of a neck node from a patient with a known primary carcinoid neoplasm in the pancreas. Discrete clusters of very cohesive uniform hyperchromatic cells with a thin rim of scanty cytoplasm are present in a clear background. Although the chromatin does not have the 'salt-and-pepper' distribution said to be characteristic of a carcinoid tumour, the overall features are compatible with the clinical diagnosis of metastatic carcinoid. ×400.

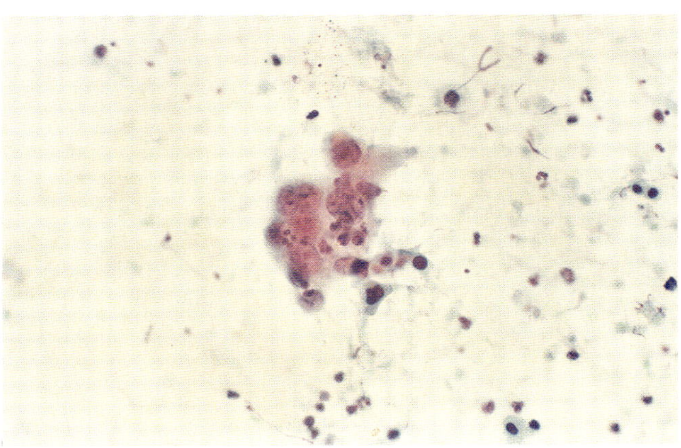

Figure 2.20 Metastatic adenocarcinoma from the stomach. A cluster of malignant pleomorphic cells with plentiful cytoplasm and large eosinophilic nucleoli is present in a background of cellular debris. The latter feature is often seen in gastric brushings from carcinoma of the stomach. ×400.

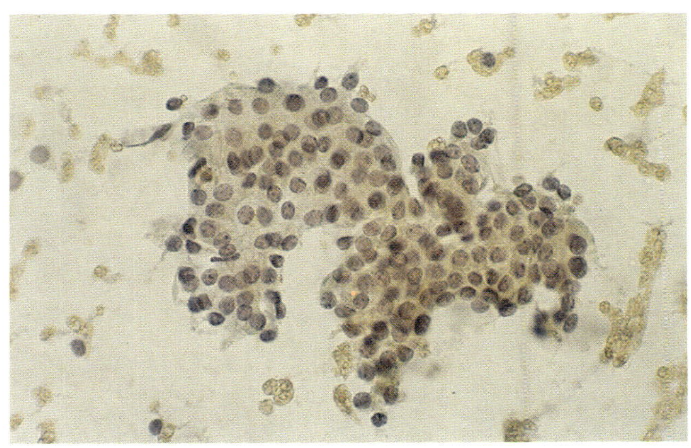

Figure 2.21 FNA of a neck node from an elderly male patient who had numerous clinical metastases. He was severely emaciated and terminally ill. The sheet of cells with a 'honeycomb' arrangement shows marked uniformity in size and shape. The nuclei are round and uniform. There are no nuclear grooves or nucleoli. Although no distinguishing features are present it was felt that this was most likely to be a metastatic adenocarcinoma, primary unknown. ×400.

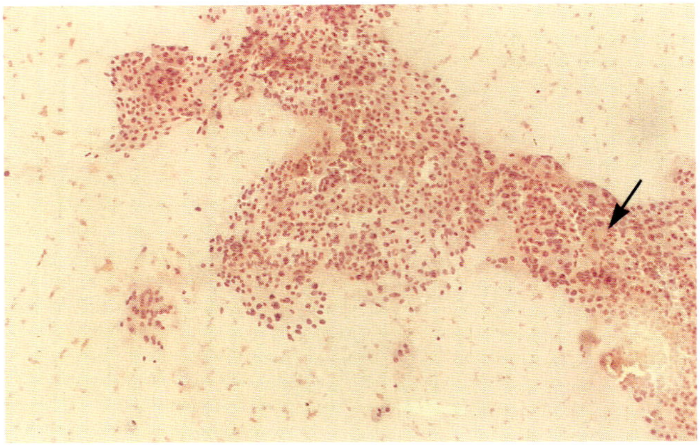

(a)

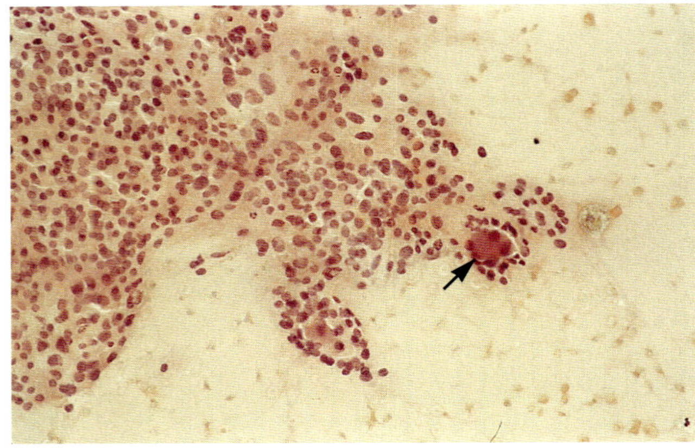

(b)

Figure 2.22 (a) and (b) Metastatic carcinoma from the thyroid. FNA of a neck node in a patient with previous history of follicular carcinoma. Sheets of uniform cells with an obvious follicular arrangement. Note clump of magenta staining colloid within a follicle (arrows). ×100.

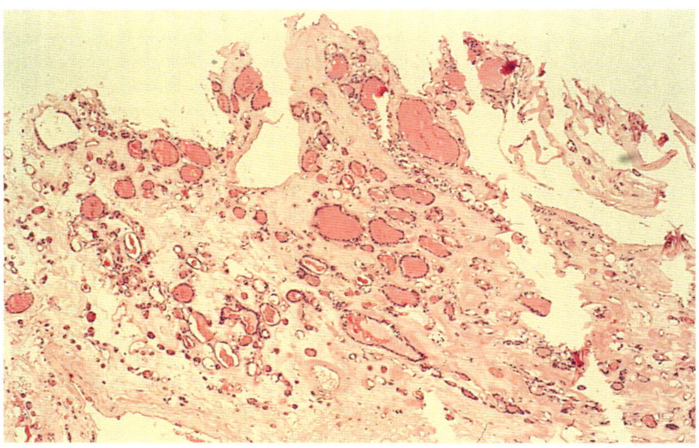

Figure 2.23 Histological section of a Trucut biopsy confirmed the presence of metastatic follicular carcinoma of the thyroid. ×100. Note: without the clinical history it is not possible to state whether thyroid tissue is 'ectopic' or metastatic thyroid carcinoma on FNA specimens.

METASTATIC ADENOCARCINOMA WITH PSAMMOMA BODIES

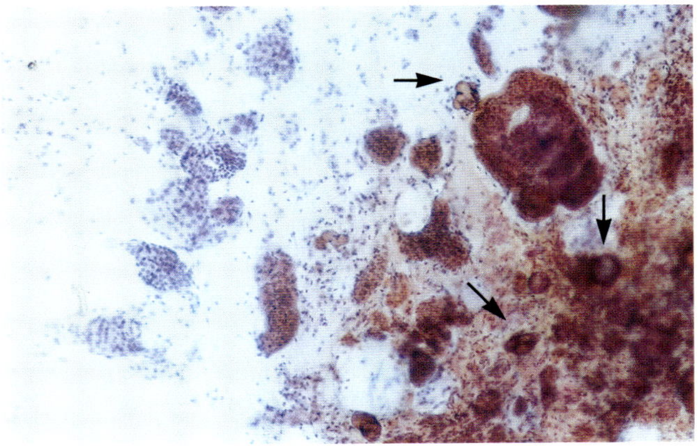

(a)

Figure 2.24 (a) FNA of a metastasis in a neck node from a primary adenocarcinoma of the palate. Numerous loose-lying glandular groups of uniform epithelial cells are present with a few psammoma bodies (arrows). These appearances are compatible with metastatic well-differentiated adenocarcinoma. ×100.

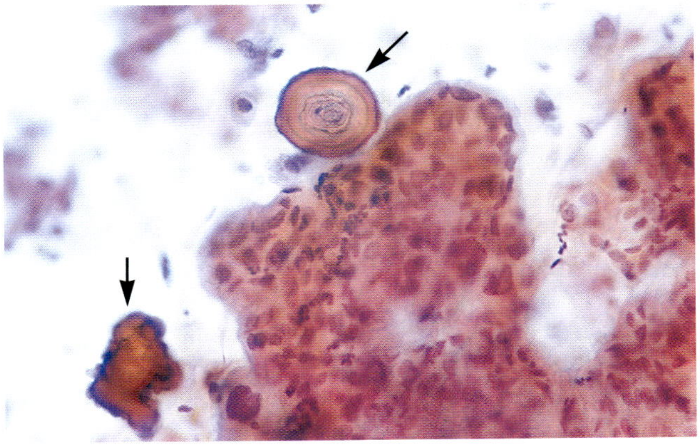

(b)

Figure 2.24 (b) High-power view of the case illustrated in (a) showing a microbiopsy of metastatic adenocarcinoma with adjacent irregular calcific body and a psammoma body. The latter exhibits concentric refractile rings. ×400.

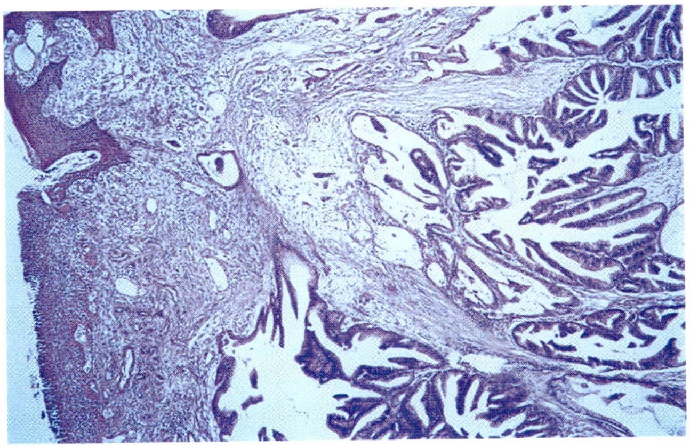

(c)

Figure 2.24 (c) Histological section of the metastatic adenocarcinoma shown in (a) and (b). Psammoma bodies and calcification are absent – they are often lost during histological processing. H&E ×100.

METASTATIC CARCINOMA WITH PROMINENT VASCULAR CHANNELS

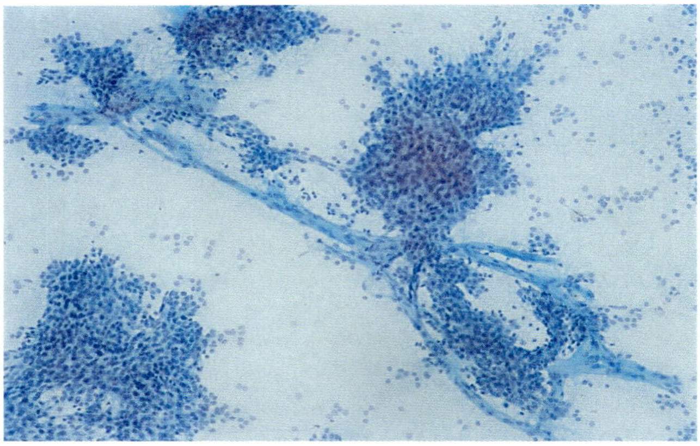

(a)

Figure 2.25 (a) Metastatic adenocarcinoma in a neck node with prominent blood vessels. This photomicrograph shows groups and clusters of uniform cells traversed by intact vascular channels. ×100.

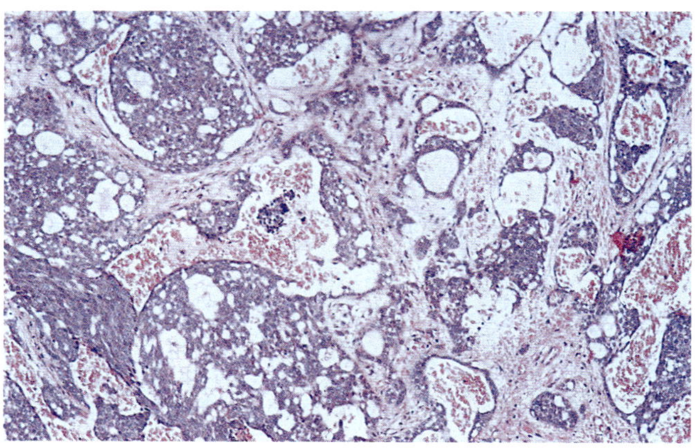

(b)

Figure 2.25 (b) Histological section of the case shown in (a), showing metastatic well-differentiated adenocarcinoma with prominent vascular channels. ×100.

METASTATIC MALIGNANT MELANOMA

It has been said that malignant melanoma is the greatest mimic in medicine. This is certainly true in cytology. First, the cells may or may not contain pigment – up to 90% of metastases do not show obvious intracellular pigment. The pigment may be coarse or dusty. It may be so plentiful as to obscure all cellular detail, or it may be scanty or absent. Secondly, the malignant melanoma cells may be obscured by lymphoid cells and thus dismissed as immunoblasts in a reactive node.

Thirdly, four different types of cells may be seen in an FNA specimen of malignant melanoma:

- small, round, lymphoid-type cells with a small rim of cytoplasm, round nuclei and prominent nucleoli
- giant multinucleated cells with plentiful cytoplasm, pleomorphic nuclei and large eosinophilic bullet-shaped nucleoli
- large binucleate cells with symmetrical eosinophilic nucleoli, giving them a similar appearance to a Reed—Sternberg cell
- spindle cells, which are usually loose lying with pale cytoplasm and oval nuclei with pinpoint nucleoli.

Finally, there may be no history of a previously excised melanotic lesion, either because enlarged neck nodes may be the first manifestation of the neoplastic spread, or because the lesion is hidden under hair on the scalp. Whenever pigment is found in an FNA specimen, careful clinicopathological correlation should be done before issuing a definitive cytopathological diagnosis.

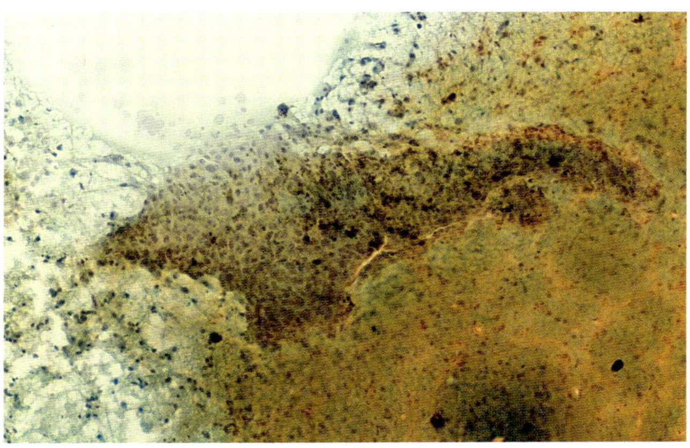

Figure 2.26 Metastatic malignant melanoma in a neck node: low-power view of a very cellular aspirate, showing a microbiopsy of very cellular tissue surrounded by loose-lying pigmented cells. The pigment is golden brown. Even at this low power the possibility of metastatic melanoma may be considered. The presence of pigment in neck nodes often poses a problem, as the neck nodes drain a very large anatomical area and thus are often the site of haemosiderin deposition after surgery, or long-forgotten trauma to the head and neck. An unusual cause of black carbon pigment deposition in neck nodes may be seen in African hut-dwellers (see Chapter 4). ×100.

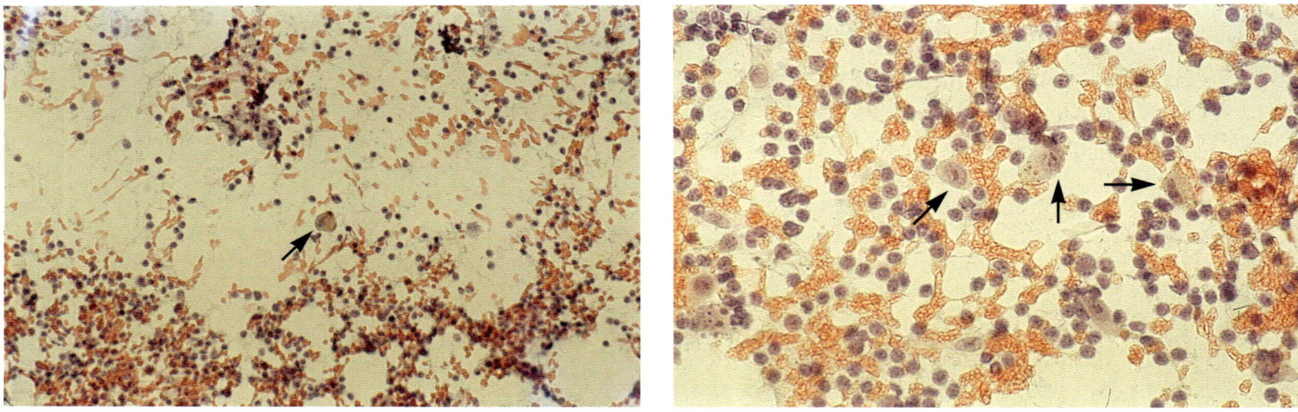

(a) (b)

Figure 2.27 (a) and (b) High-power view of malignant melanoma cells (arrows), intermixed very subtly with lymphoid cells. They are loose lying and pleomorphic. Note the irregular macronucleoli. Some contain golden-brown pigment, which is dusty in some cells and granular in others. ×400.

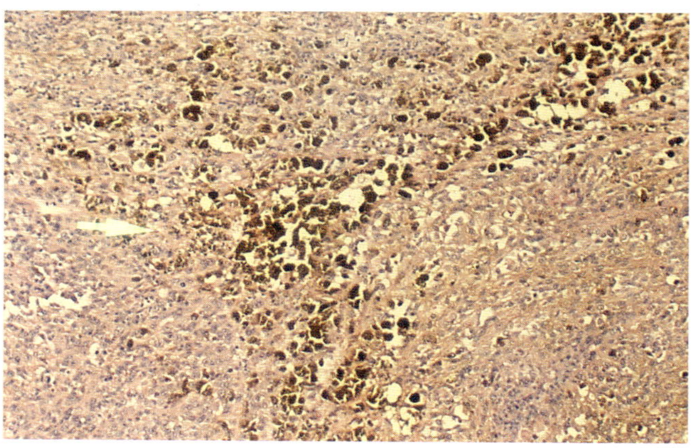

Figure 2.28 Histology of the neck node confirmed the presence of a pigmented metastatic malignant melanoma. H&E ×100.

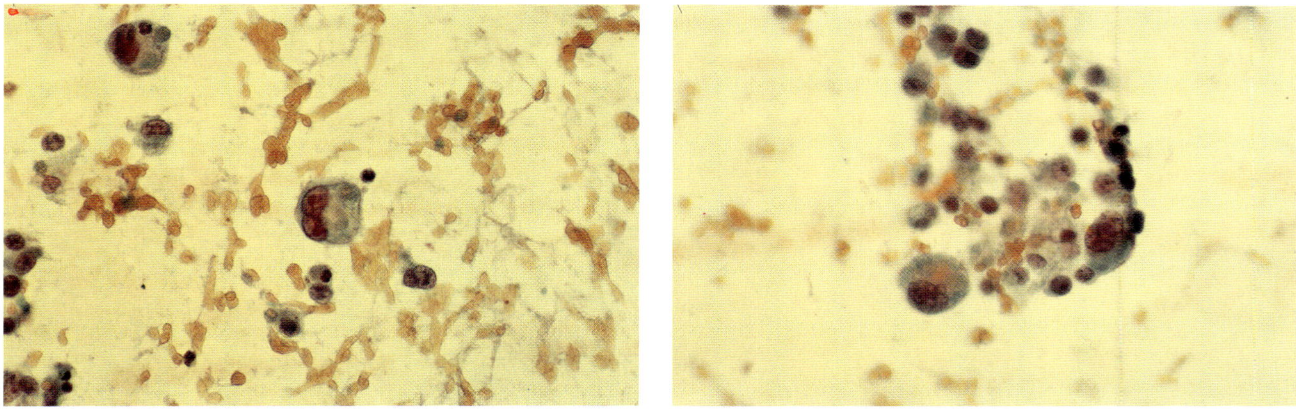

(a) (b)

Figure 2.29 (a) and (b) Metastatic malignant melanoma: FNA of a neck node in a female patient with multiple subcutaneous nodules all over the head, neck and trunk, and a mass in the breast. There was no history of removal of a pigmented lesion. There are scattered loose-lying cells of varying sizes, some containing fine dusty pigment. There are small lymphocyte-like cells with a rim of cytoplasm. They have round regular nuclei and prominent large nucleoli. Large binucleate cells with large prominent nucleoli are present. Although these are reminiscent of Reed—Sternberg cells, they have very plentiful cytoplasm and there are no lymphoid cells in the background. ×400.

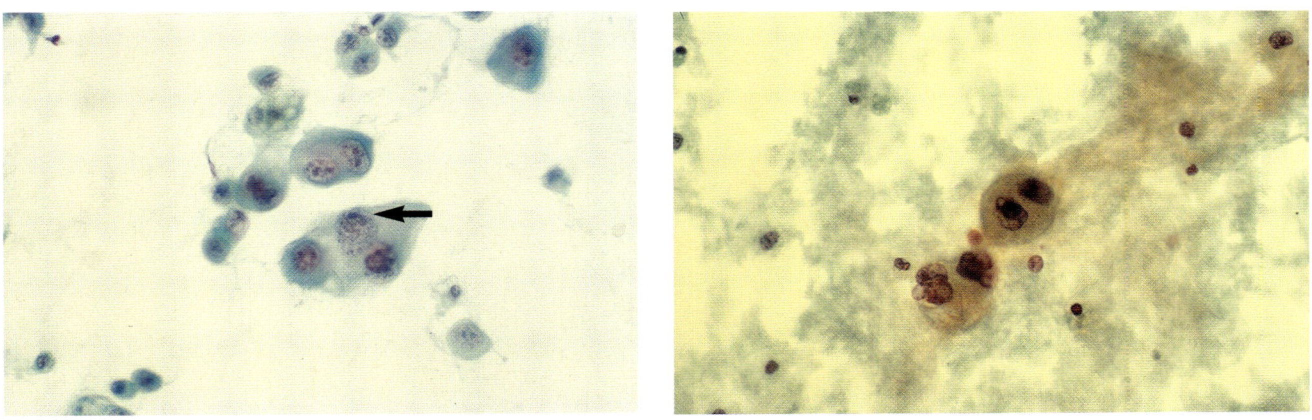

(a) (b)

Figure 2.30 (a) and (b) Simultaneous FNAs from the axilla and breast showed loose-lying large binucleate pigmented cells with macronucleoli compatible with metastatic malignant melanoma, primary site unknown. ×400.

METASTATIC OLFACTORY NEUROBLASTOMA

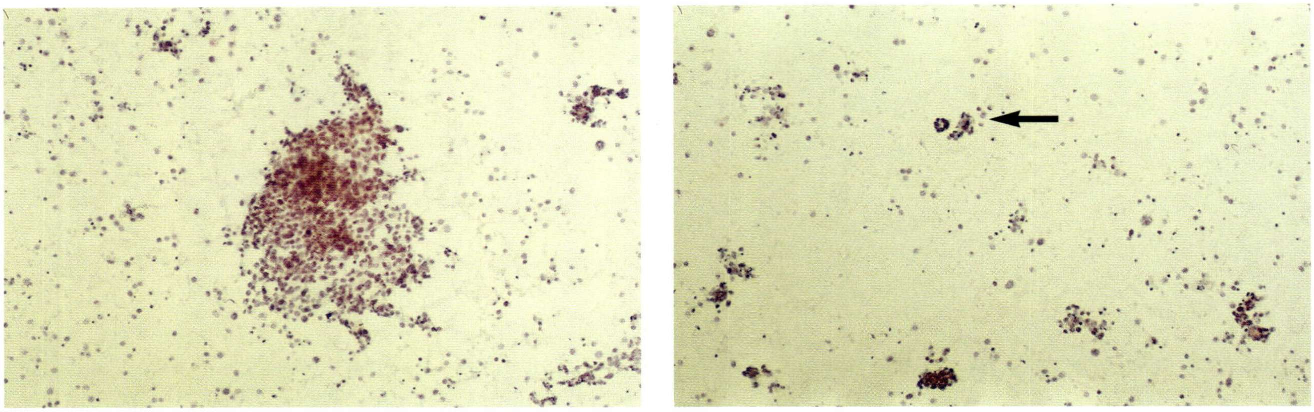

(a) (b)

Figure 2.31 (a) and (b) Metastatic olfactory neuroblastoma or neuroendocrine tumour. This child had a known small round-cell neoplasm in the nasal region, for which he had been treated. FNA of an enlarged neck node was done. A very cellular aspirate composed of loose clusters of small, round, apparently uniform blast-like cells was seen. An attempt at rosette formation can be discerned at low power. ×100.

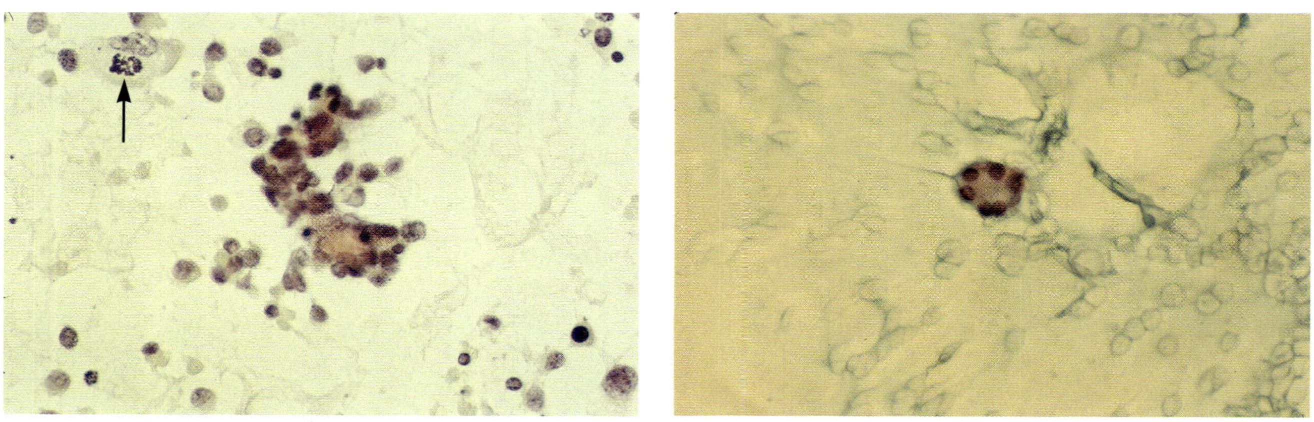

(a) (b)

Figure 2.32 (a) and (b) High power shows rosette formation. The cells are more pleomorphic than is usual in this type of neoplasm. An abnormal mitotic figure is noted (arrow). ×400.

REFERENCES AND FURTHER READING

Carson, H.L., Candel, A.G., Gattuso, P. and Castelli, M.J. (1996) Fine-needle aspiration of supraclavicular lymph nodes. *Diagn. Cytopathol.* **14**: 216–220.

Howell, L.P., Russell, L.A., Howard, P.H. and Teplitz, R.L. (1992) The cytology of paediatric masses: a differential diagnostic approach. *Diagn. Cytopathol.* **8**: 107–115.

Miller, T.R., Bottles, K., Abele, J.S. and Beckstead, J.H. (1985) Neuroblastoma diagnosed by fine-needle aspiration biopsy. *Acta Cytol.* **29**: 461–468.

Schwarz, R., Chan, N.H. and Macfarlane, J.K. (1990) Fine-needle aspiration cytology in the evaluation of head and neck masses. *Am. J. Surg.* **159**: 482–485.

Steel, B.L., Schwartz, M.R. and Ramzy, I. (1995) Fine-needle aspiration biopsy in the diagnosis of lymphadenopathy in 1103 patients. Role, limitations and analysis of diagnostic pitfalls. *Acta Cytol.* **39**: 76–81.

Strickler, J.G., Roger, A.W. and Weiss, L.M. (1987) Necrosis in lymph nodes. *Path. Ann.* **II**: 253–282.

Theunnissen, F.B.J.M., Peterse, J.L., van der Pel, R. *et al.* (1993) Reliability of fine-needle aspiration cytology for distinguishing between carcinoma, lymphoma and sarcoma: the influence of clinical information. *Cytopathology* **4**: 107–114.

3 Cysts of the head and neck

Soft fluctuant masses in the head and neck are often diagnosed clinically as cysts. Layfield *et al.* (1987) reported that cystic lesions are a common source of erroneous diagnosis. They also reported cystic change occurring in pleomorphic adenoma (PA), adenoid cystic carcinoma, Warthin's tumour and chronic sialadenitis. Central necrosis in metastatic squamous carcinoma is common and often gives rise to an incorrect clinical diagnosis of a cystic mass. It is not possible to distinguish between a developmental cyst (branchial cyst) and a metastatic carcinoma on cytological evidence alone (Engzell and Zajicek, 1970). Clinicopathological correlation, particularly the age of the patient, is of paramount importance when reporting on cystic masses. In an area where there is endemic tuberculosis, this diagnosis must always be considered when dealing with apparently 'cystic' masses. Where parasitic disease is common, strange and unusual 'dust contaminants' should always be given particular attention.

PLEOMORPHIC ADENOMA WITH CYSTIC CONSISTENCY

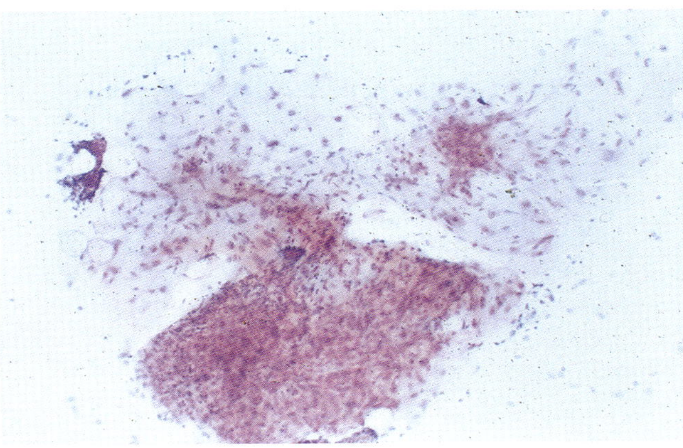

Figure 3.1 FNA of this 'soft cystic' mass in the parotid region yielded thick 'mucoid' material. However, the smears revealed a very cellular aspirate with two distinct components: groups of tightly cohesive epithelial cells which are uniform in size and shape, and fragments of metachromatic myxoid (almost chondroid) tissue, which is very cellular. The cells show some nuclear pleomorphism. These features are suggestive of a pleomorphic adenoma with a predominant myoepitheliomatous component. These variants of PA may be very fluctuant on palpation and may be mistaken clinically for a cyst. ×100.

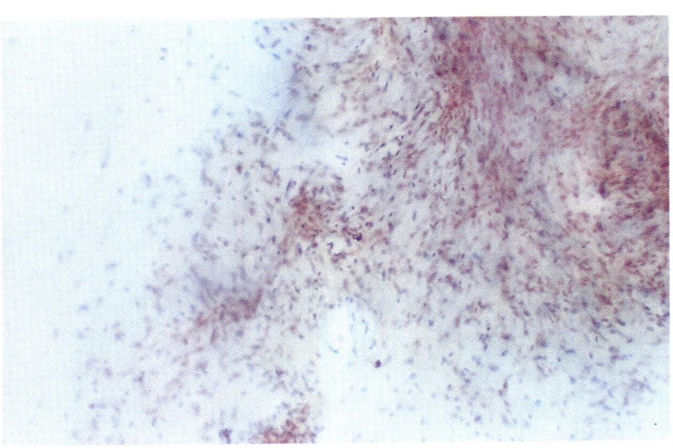

Figure 3.2 Another view of the case illustrated in Figure 3.1 shows a microbiopsy of loose, cellular stromal tissue suggestive of a myoepitheliomatous neoplasm. No epithelial cells are seen. ×400.

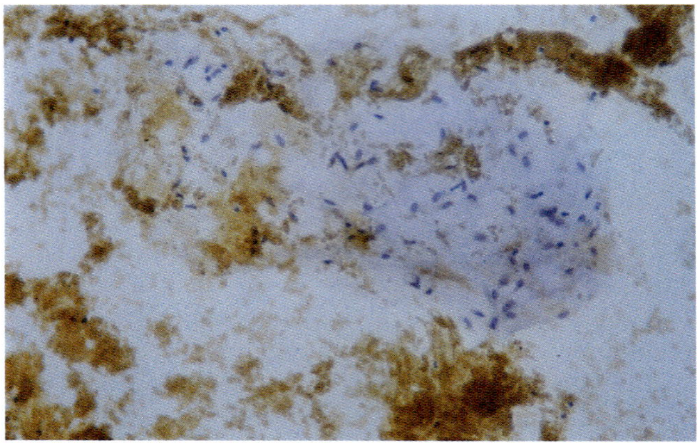

Figure 3.3 FNA of a soft cystic mass below the angle of the mandible in an elderly patient showing fragments of loose, almost transparent but cellular tissue. No epithelial-like cells were found. A soft tissue neoplasm was cytologically suspected, e.g. myxoma, lipoma, fibromyxoma; however, on histology a pleomorphic adenoma was found. The parotid glands are known to descend into the neck with age and may therefore be mistaken for metastatic nodes. ×400.

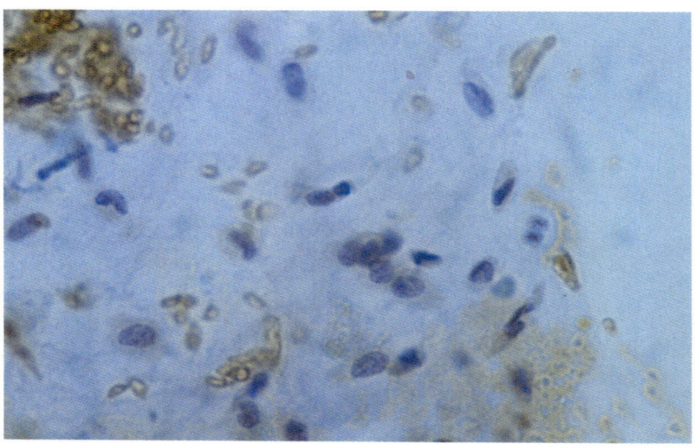

Figure 3.4 High-power view of the stromal cells and nuclei from the specimen shown in Figure 3.3. ×400.

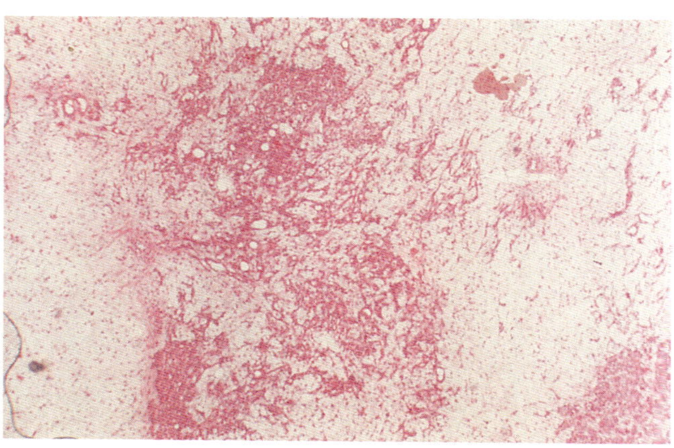

Figure 3.5 Histological section of a pleomorphic adenoma with predominant myoepitheliomatous component. The problem of inaccurate classification of parotid neoplasms on FNA specimens has been reviewed by Orell and Nettle (1988). H&E ×100.

DEVELOPMENTAL CYSTS

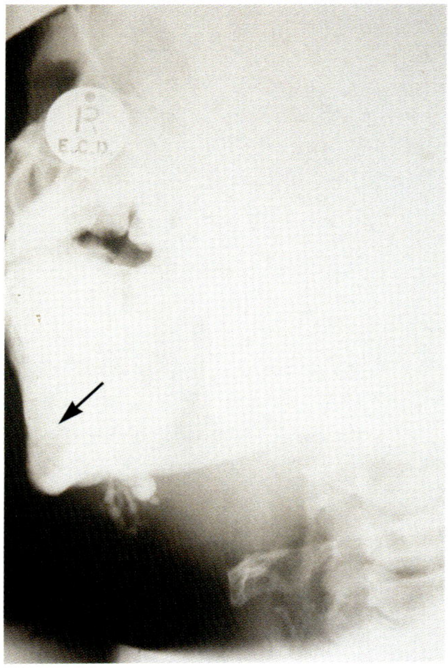

Figure 3.6 X-ray of cystic lesion in the mandible of a 25-year-old male causing a lytic lesion of bone, enabling an FNA to be done. Clear fluid was aspirated (see Figure 3.7).

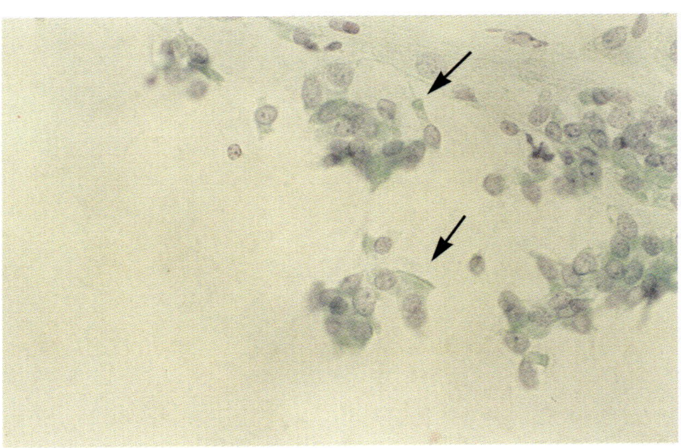

Figure 3.7 Very scanty, loose-lying columnar epithelial cells with basal nuclei are present in a clean background. Terminal bars and degenerate cilia are seen (arrows). No squamous cells or necrotic debris are present. These features are very suggestive of a developmental fissural cyst. ×100.

BRANCHIAL CYSTS

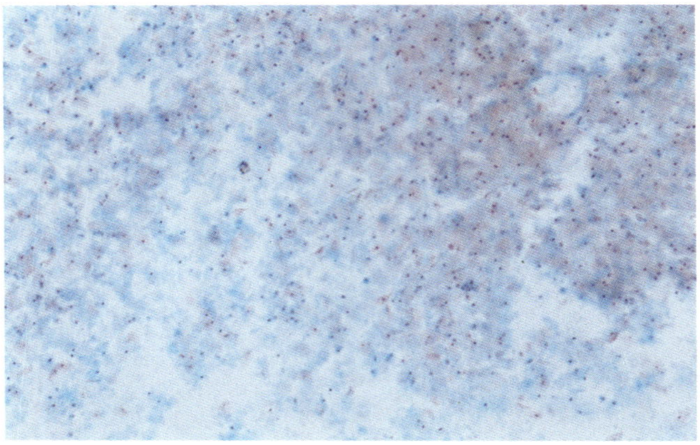

(a)

Figure 3.8 (a) FNA of a cystic neck mass in a teenaged patient. Low power view shows loose lying benign anucleated and nucleated squames in a clean background.

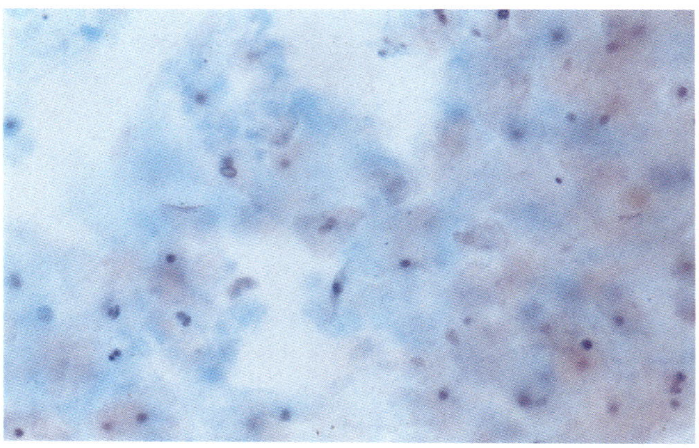

(b)

Figure 3.8 (b) High power view emphasizes the benign nature of these squamous cells. The nuclei are small, round and centrally placed within the cells.

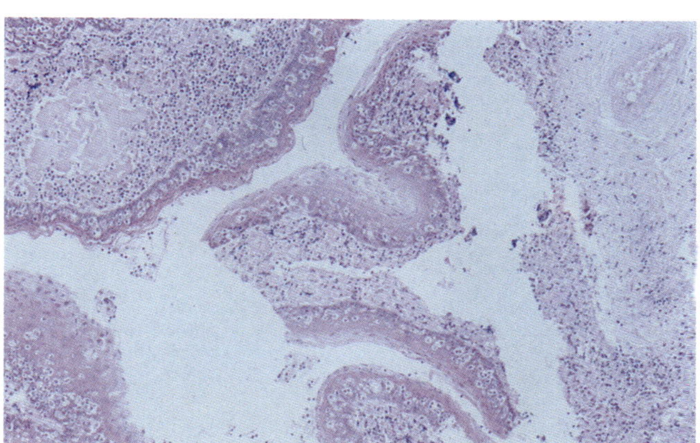

(c)

Figure 3.8 (c) Histological section of the cystic mass discussed in Figures 3.8 (a) and (b) shows cystic spaces lined by squamous epithelium.

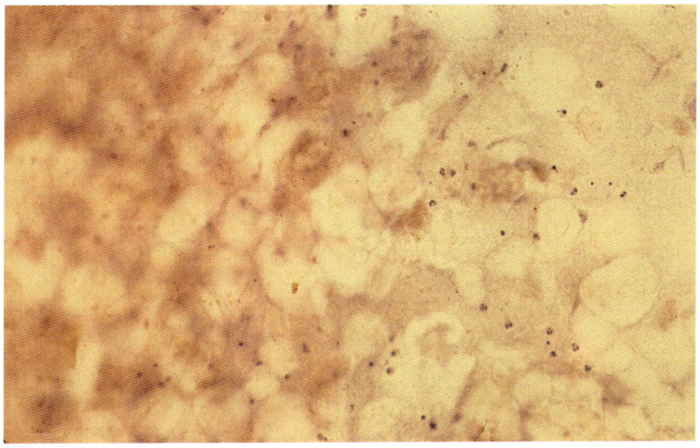

Figure 3.9 FNA of a branchial cyst. High-power view of anucleate squames with melanin pigment. Note faint 'basketweave' appearance of the cellular material. ×400.

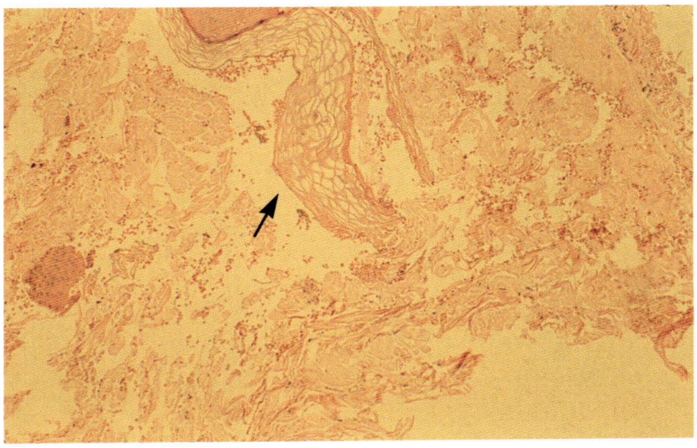

Figure 3.10 This view of a histological section of the resected cyst discussed in Figure 3.9 shows fragments of anucleate squamous epithelium with a basketweave appearance (arrow), lying in a background of debris. Note brown melanin pigment. ×100.

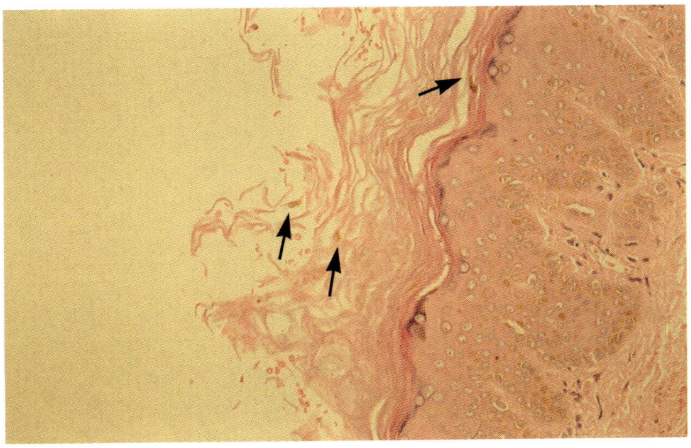

Figure 3.11 Histology of the cyst wall shows the thickened basketweave squamous epithelium containing pigmented cells (arrow), as seen in the cytological preparation. (Figure 3.9) ×400.

CYSTIC CHANGE IN METASTATIC NECK NODES

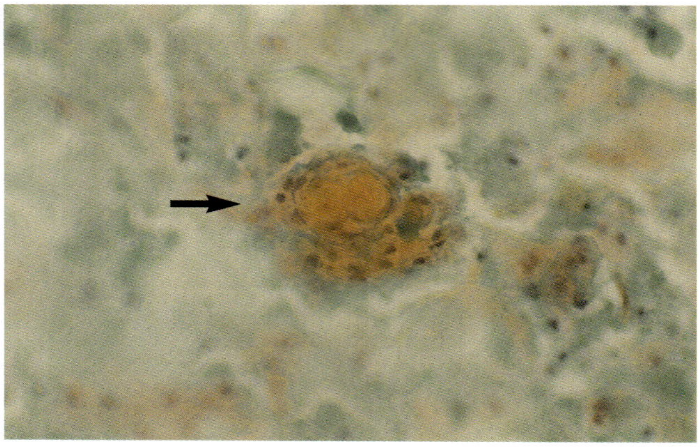

Figure 3.12 FNA of a recurrent subcutaneous cystic swelling in the posterior triangle of the neck of a 70-year-old male. The aspirate was presumed to be pus, but a portion of it was nevertheless submitted for cytological examination. Loose-lying, highly atypical squamous cells and a squamous pearl are seen in a background of necrotic material. The differential diagnosis is (a) metastatic squamous carcinoma; (b) a carcinoma arising in a pre-existing branchial cyst; or (c) extension from a primary squamous carcinoma of the skin. ×400.

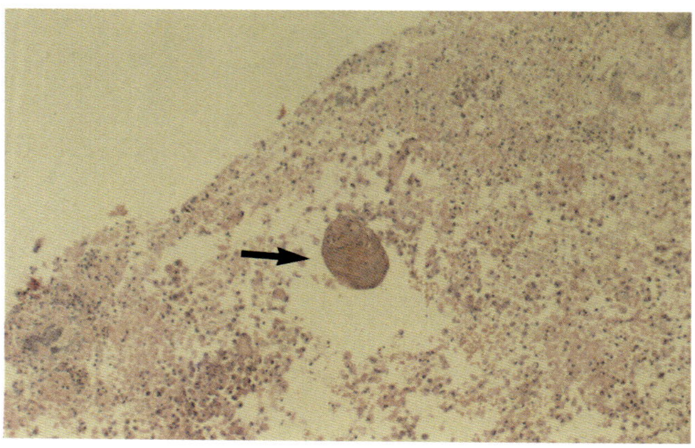

Figure 3.13 Histology of the case shown in Figure 3.12 shows a squamous pearl lying in a background of cellular debris with adjacent squamous carcinoma. H&E ×100.

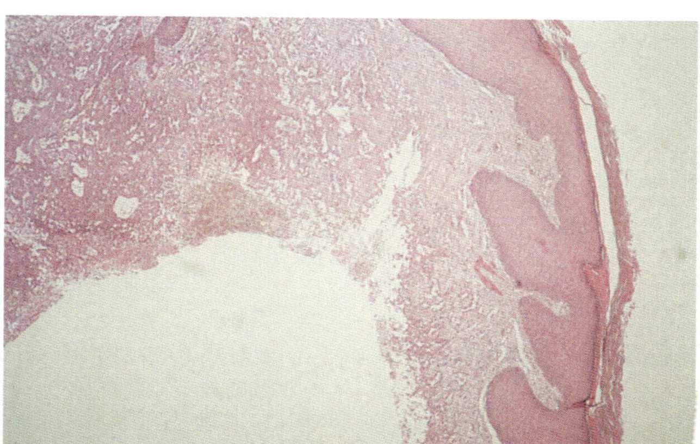

Figure 3.14 This histological view of the subcutaneous mass discussed in Figures 3.12 and 3.13 shows the cystic nature of the metastatic mass, with islands of invasive squamous carcinoma lying in a necrotic background. H&E ×100.

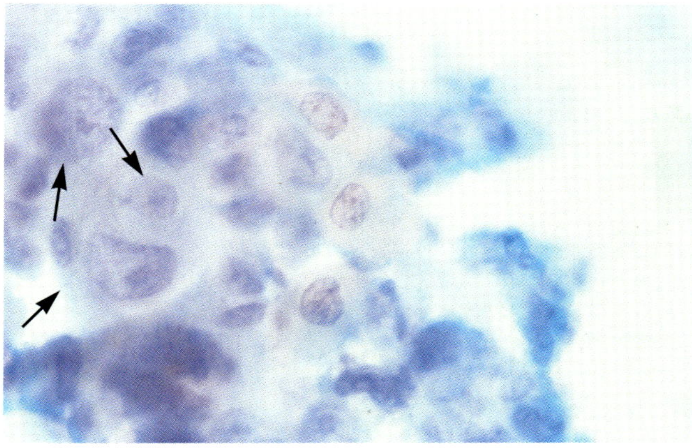

Figure 3.15 FNA of a submandibular cystic mass in a patient with a known history of squamous carcinoma of the lip. A sheet of highly atypical pleomorphic cells with plentiful basophilic cytoplasm and large irregular nuclei with disturbed chromatin and prominent macronucleoli is seen (arrows). These features are consistent with a metastatic squamous carcinoma. ×100.

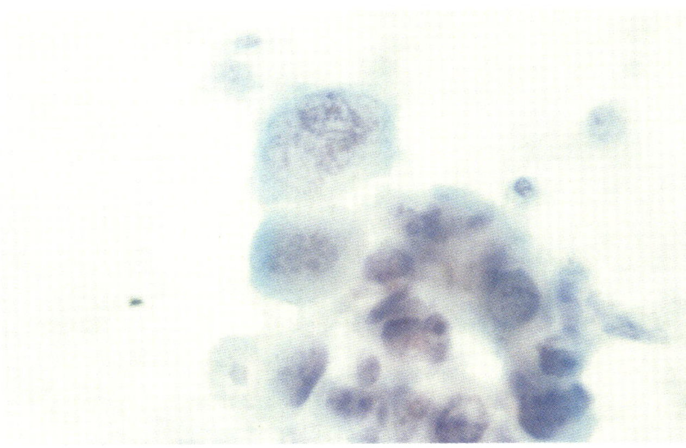

Figure 3.16 Metastatic squamous carcinoma in a "cystic" submandibular node. Large pleomorphic basophilic cells are shown. The nuclear chromatin is markedly disturbed and uneven. An abnormal mitotic figure is present. The patient had had radiotherapy. ×400.

TUBERCULOSIS IN A CYSTIC NODE

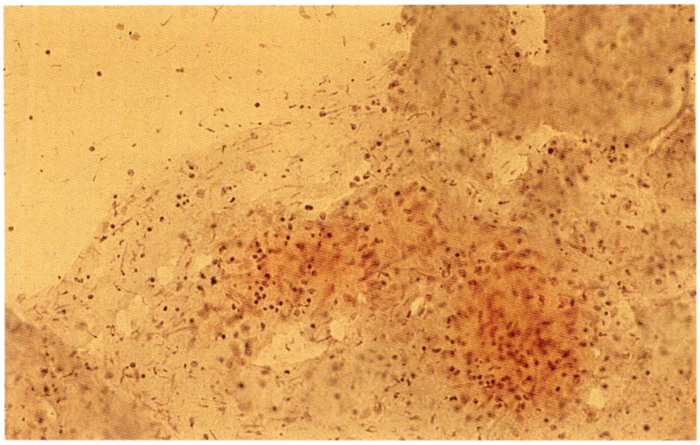

Figure 3.17 FNA of a fluctuant mass in the neck, thought clinically to be a cyst. A loose granulomatous-like structure is present in a background of amorphous material containing wisps of chromatin. Many of the latter are shaped like a comma and are often seen in FNA material from tuberculous masses (Gupta *et al.*, 1992; Rajwanshi *et al.*, 1987). The granuloma is composed of pale, uniform, oval-shaped epithelioid cells intermixed with lymphocytes. No Langhans' giant cells are present in this specimen, which probably represents a case of acute tuberculosis (tuberculosis was epidemic in the patient's residential area) (see Chapter 4). ×400.

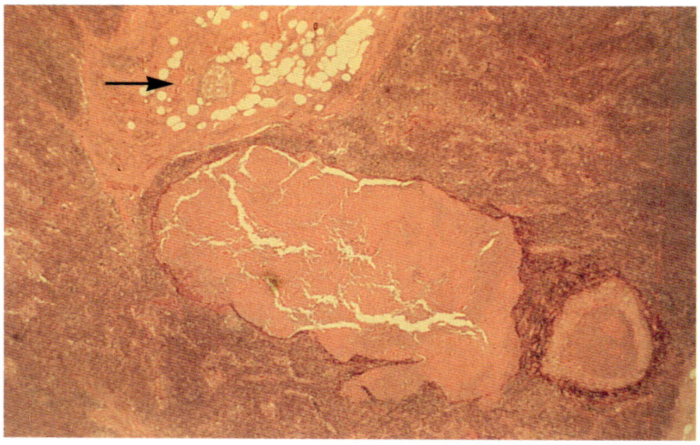

Figure 3.18 Histological section of the mass shown in Figure 3.17 shows widespread destruction of lymphoid tissue by caseation necrosis, compatible with acute tuberculosis. A few small loose granulomata were found (arrow). H&E ×400.

SEBACEOUS ADENOMA/HYPERPLASIA

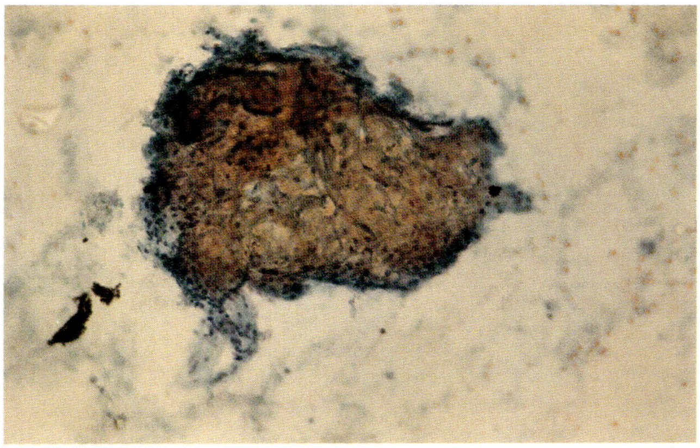

Figure 3.19 FNA of a soft cystic mass on the cheek of a 75-year-old male. Cohesive clusters of round cells with well-defined cytoplasmic membranes were present, suggestive of an adnexal skin tumour, possibly sebaceous in origin. These cells could represent sebaceous hyperplasia or sebaceous adenoma. ×400.

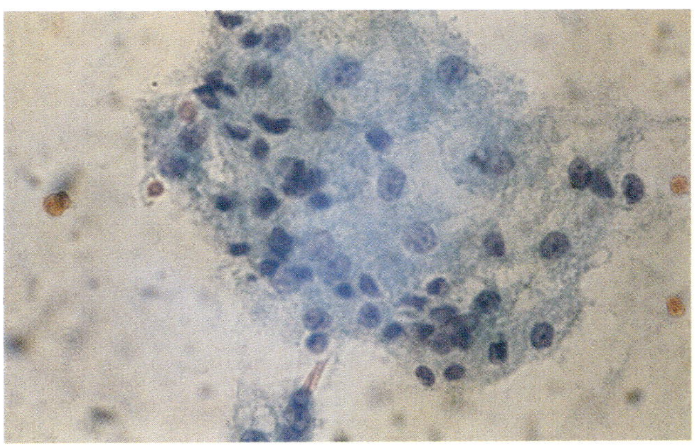

Figure 3.20 A cluster of sebaceous cells showing clear 'bubbly' cytoplasm, pale central nucleoli and pinpoint nucleoli. ×100.

ECHINOCOCCUS GRANULOSUS IN A 'CYSTIC NODE'

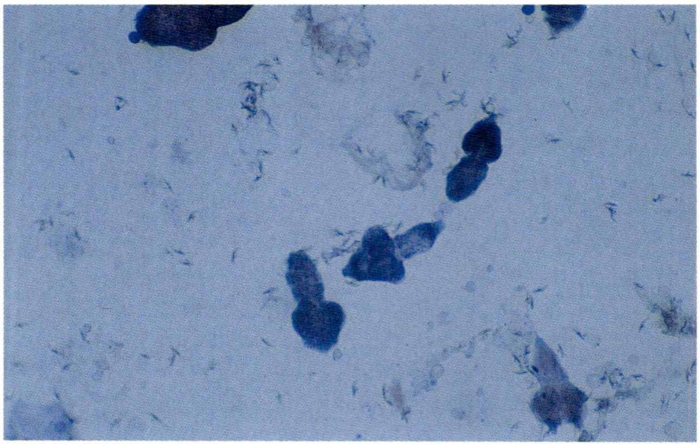

Figure 3.21 FNA of a soft, cystic, painless mass in the parotid gland of a 56-year-old female. It had fluctuated in size from 3 to 5 cm. A moderate amount of clear fluid was aspirated. It was initially regarded as acellular by the cytotechnologist and a junior resident; however, a final check by the author revealed that the clumps of 'contaminant dust' were in fact scolices and attached segments of a parasitic tapeworm. ×100.

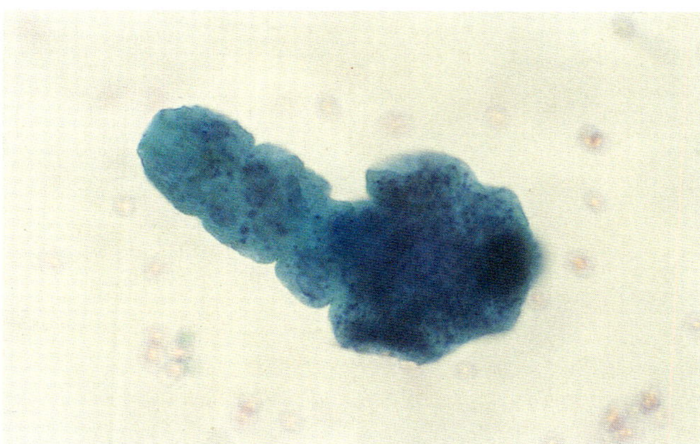

Figure 3.22 Easily recognized scolex, suckers and attached segments of *Echinococcus granulosus*. ×400.

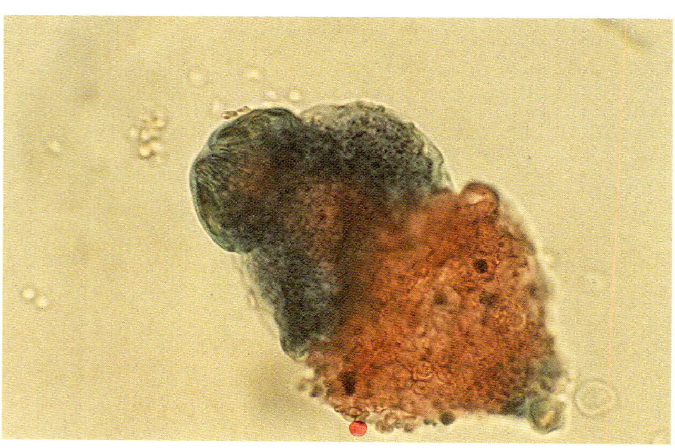

Figure 3.23 Note the refractile nature of the rostellum of hooklets on the 'crown' of the scolex, giving it the semblance of a flower head. The attached coarse granular eosinophilic material composed of spherules of calcific material is known as 'hydatid sand'. ×400.

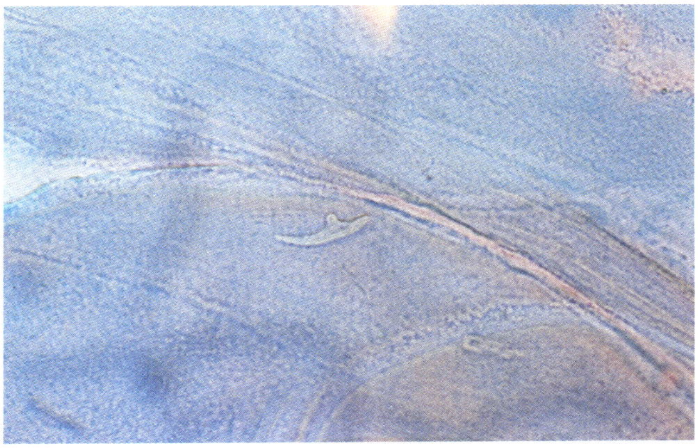

Figure 3.24 High-power view of single-lying hooklet. These can easily be missed on screening unless one is vigilant when looking at fine-needle aspirates composed of clear fluid. Oil immersion. ×1000

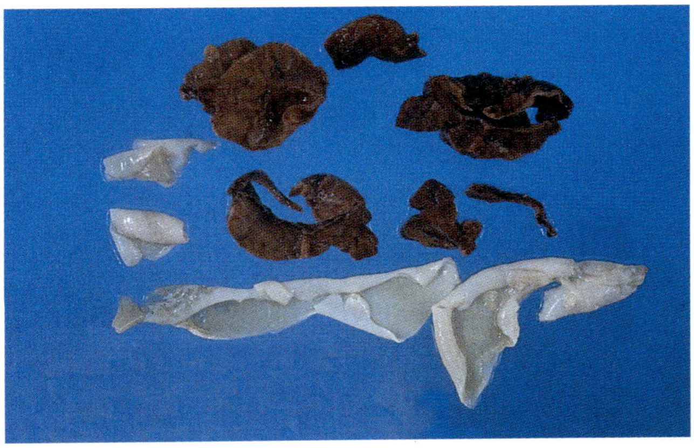

Figure 3.25 Resected specimen showing fragments of white cyst wall of *Echinococcus granulosus* and remnants of salivary gland tissue shown in Figures 3.21–3.24. The cyst, which was intimately adherent to the salivary gland, ruptured during surgical removal. A Cytospin preparation of the formalin fluid from the specimen jar also showed many scolices and hooklets (see Chapter 10).

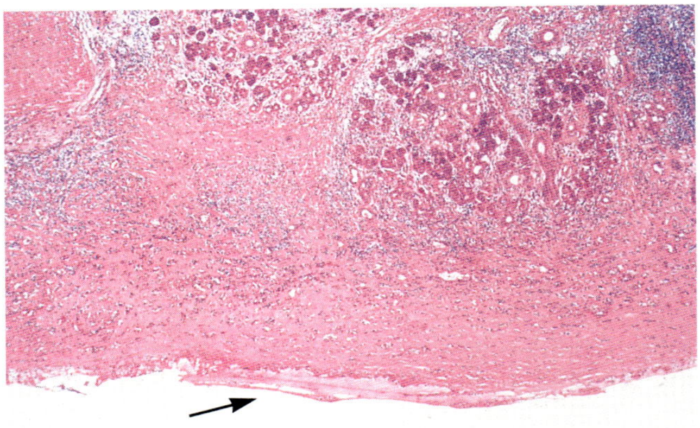

Figure 3.26 Histological section of the wall of the hydatid cyst shown in Figure 3.25 shows granulation tissue surmounted by a thin, laminated cyst membrane (arrow), with adjacent salivary gland tissue. H&E ×100.

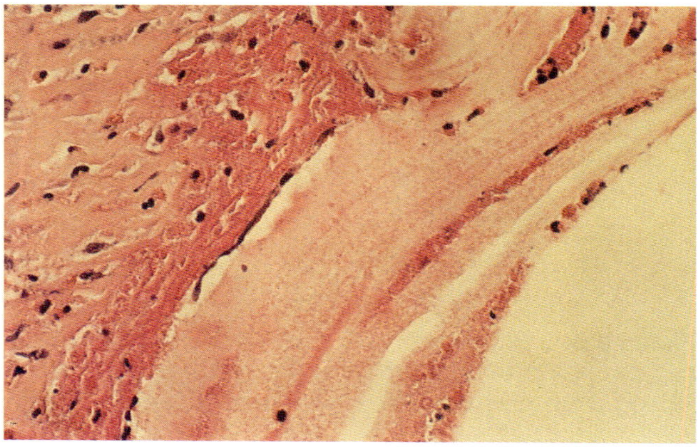

Figure 3.27 High-power view of the laminated membrane of the hydatid cyst wall with a layer of fibrosis intervening between it and the salivary gland tissue. H&E ×400.

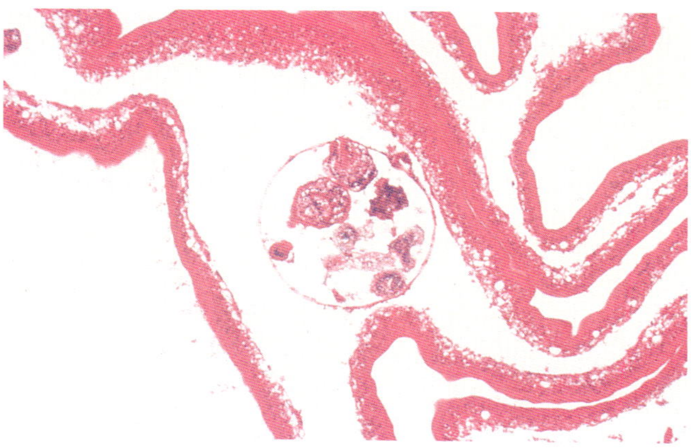

Figure 3.28 Histological appearance of the hydatid cyst shows layers of hydatid cyst wall with daughter cysts incorporating scolices. ×100.

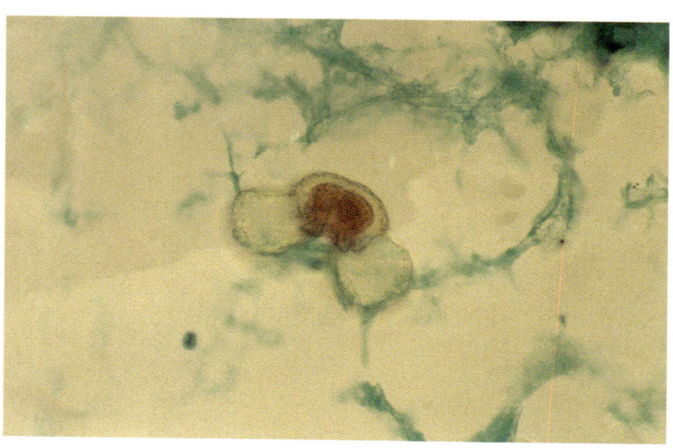

Figure 3.29 This structure was noted on another FNA of a parotid gland. It resembled a parasite and was mistaken for one by a junior cytotechnologist. It is in fact a pollen grain from the flora on the slopes of Table Mountain. These are often mistaken for organisms by the unwary. ×400.

INFLAMMATORY CYSTIC MASS

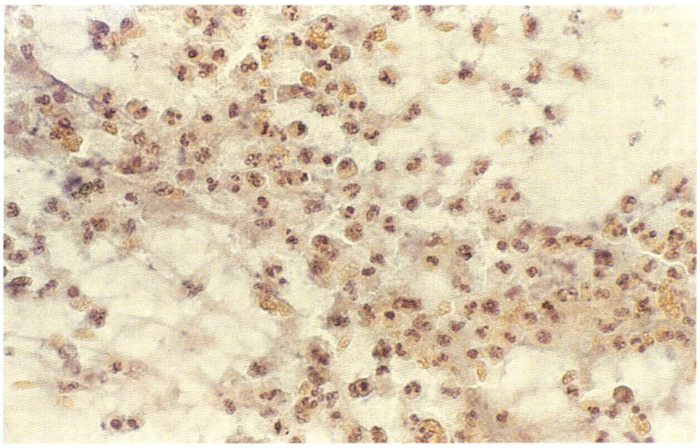

Figure 3.30 FNA of a 'cystic mass of the larynx' yielded mucopurulent material with numerous polymorphs and debris. No epithelial cells, malignant cells or bacteria or fungi were seen. ×400.

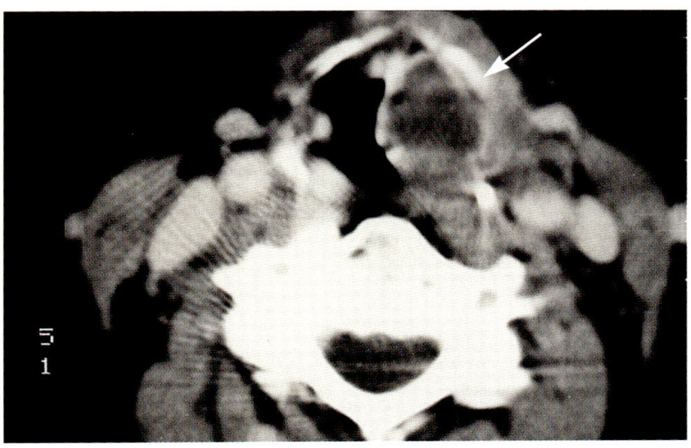

Figure 3.31 Review of the laryngeal CT scan showed a cystic lesion which was considered to be consistent with a laryngopyelocoele (arrow).

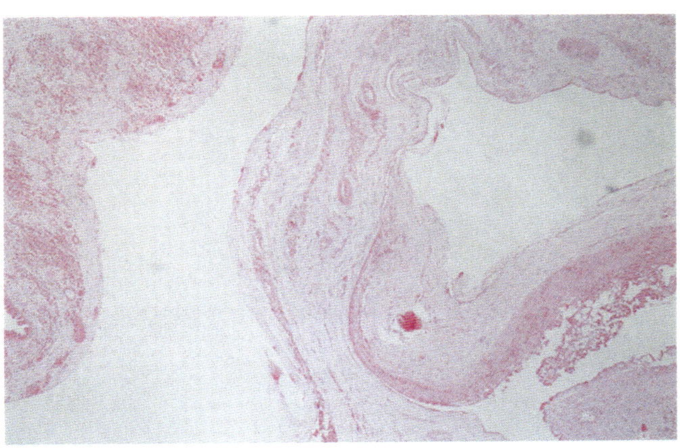

Figure 3.32 The lesion shown in Fig. 3.30–3.31 was successfully excised. Histological sections of the wall of the cystic mass which was lined by metaplastic cells and columnar cells, failed to reveal any dysplasia/malignant change. H&E ×100.

REFERENCES AND FURTHER READING

Borghese, M. R. and Romanelli, L. (1994) Intraoperative cytology in otorhinolaryngological and cervical pathology. *Pathologica* **86**: 291–296.

Engzell, U. and Zajicek, J. (1970) Aspiration biopsy of tumours of the neck. I. Aspiration biopsy and cytologic findings in 100 cases of congenital cysts. *Acta Cytol.* **14**: 51–57.

Gupta, A. K., Nayat, M. and Chandra M. (1992) Critical appraisal of fine-needle aspirate cytology in tuberculous lymphadenitis. *Acta Cytol.* **36**: 391–394.

Layfield, L. J., Tan, P. and Glasgow, B. J. (1987) Fine-needle aspiration of salivary gland lesions. Comparison with frozen sections and histologic findings. *Arch. Pathol. Lab. Med.* **111**: 346–353.

Orell, S. R. and Nettle, W. J. S. (1988) Fine-needle aspiration biopsy of salivary gland tumours. *Pathology* **20**: 332–337.

Palma di, S. (1996) FNA of myoepithelioma of salivary gland. *Cytopathology* **7**: 357–365.

Rajwanshi, A., Bhambani, S. and Das, D. K. (1987) Fine-needle aspiration cytology diagnosis of tuberculosis. *Diagn. Cytopathol.* **3**: 13–16.

Verma K, and Mandal, S., Kapila. (1995) Cystic change in metastatic squamous carcinoma. *Acta Cytol.* **39**: 478–480.

4 Lymph nodes – benign lesions

In clinical practice it is important to be able to assist the clinician in performing triage of patients with enlarged lymph nodes in the neck. Cytopathological diagnosis on an outpatient basis enables the clinician and the cytopathologist to categorize the disease process, i.e. infection, non-neoplasia, benign neoplasia, or a malignant neoplasm.

TUBERCULOSIS

Tuberculosis has become an important public health problem in many countries. The emergence of drug-resistant tuberculosis (DRT) has added to the increasing number of cases in many communities, as has the increasing incidence of HIV/AIDS.

In the periurban area of Cape Town, South Africa, the number of notified cases of tuberculosis was 26 741 in 1994, 25 636 in 1995 and 22 636 in 1996: of these, 913, 1003 and 846 died in each respective year. These statistics emphasize the need for on-site diagnosis of tuberculosis to avoid time-consuming investigations and to facilitate the immediate commencement of treatment using the World Health Organization tuberculosis control strategy, which is termed DOTS (Directly Observed Treatment, Short-course). In South Africa, patients are treated in community-based clinics.

FNA of tuberculous neck nodes is a reliable, quick and inexpensive method of diagnosis (Finfer *et al.*, 1991; Shariff and Thomas, 1991; Gupta *et al.*, 1992). Indeed, in the management and control of TB in epidemic areas it is an eminently suitable diagnostic tool. Slides are prepared in the conventional way; an extra slide can be stained with Ziehl–Neelsen stain for tubercle bacilli (see Figure 4.6), and material can be sent for culture and research if necessary. Traditionally the identification of organisms has been the domain of bacteriologists; however, cytologists have been identifying organisms for years in sputum specimens and on cervical smears. In primary health-care settings it may be more practical to have cytologists identifying tubercle bacilli (Kupper *et al.*, 1995). In epidemic areas, DOTS treatment is begun as soon as any microscopic evidence for TB is found.

Specimens obtained from tuberculous masses show a variety of cytological pictures, depending on the stage of the disease (Bailey *et al.*, 1985):

- widespread caseous necrosis without epithelioid cells, granulomata or giant cells. In this instance a Ziehl–Neelsen stain may be helpful, or a repeat aspiration can be suggested. Caseous material has a characteristic amorphous appearance. It contains wisps of chromatin, which have been likened to the shape of a comma (Koss, 1984). Mature lymphocytes may or may not be present
- epithelioid cells may be round to oval in shape or, in more mature granulomata, they are all spindle shaped (Gupta). The nuclei are pale with even chromatin. Epithelioid cells may be loosely arranged, splayed out or tightly clustered
- multinucleated giant cells are formed when epithelioid cells merge together. These are termed Langhans' giant cells (LGC). On histological sections the nuclei appear to be peripherally placed near the cytoplasmic membrane; however, on cytoplasmic preparations they may appear to be arranged more randomly
- microbiopsies of granulomatous tissue may be aspirated once the granulomata have coalesced and are merged together by fibrocollagenous tissue (Brown and Coghill, 1992). Discrete granulomata may be aspirated with or without accompanying caseous material or LGC, or lymphocytes
- occasionally a blood vessel is observed protruding from the centre or edge of a granuloma.

A high index of suspicion and heightened awareness of TB, especially in an epidemic area, assists the cytopathologist in many cases.

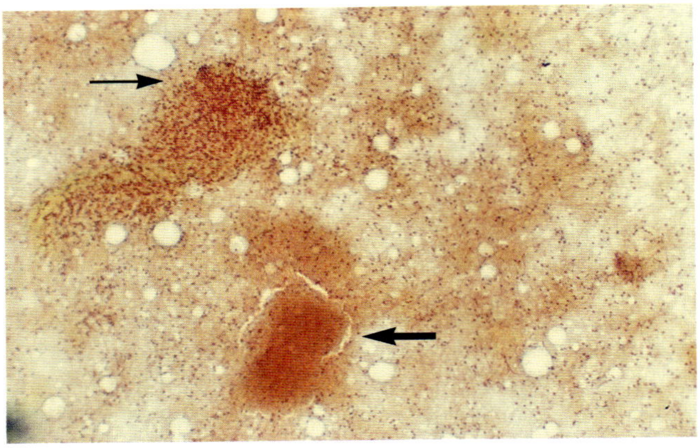

(a)

Figure 4.1 (a) FNA from an enlarged neck node in a young adult with recent rapid weight loss. This low-power view shows caseous necrotic material incorporating scattered lymphocytes and a large granulomatous-like structure surrounded by lymphocytes. A thick clump of necrotic material is present (arrow) with an artefactual clear rim around it. ×100.

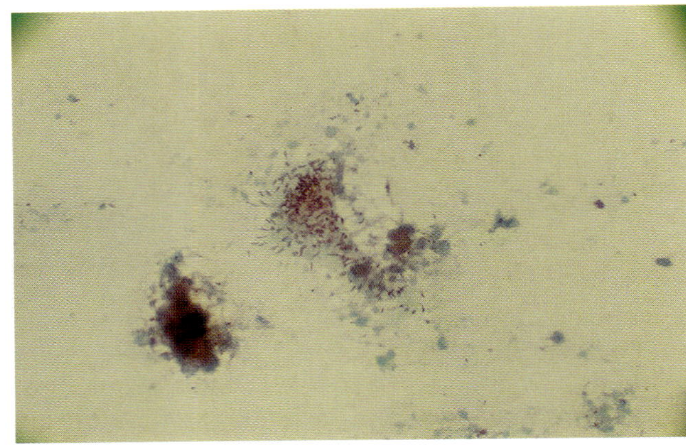

(b)

Figure 4.1 (b) Low-power view of a discrete granuloma, composed of loose, splayed out oval-shaped epithelioid cells. Small and large clumps of acellular necrotic material are also present. These appearances are compatible with the clinical diagnosis of tuberculosis. No lymphoid cells were noted – this may indicate that the patient is immunosuppressed. ×100.

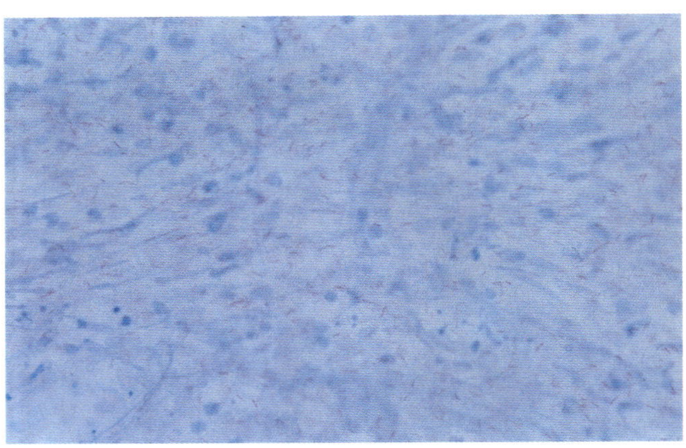

(c)

Figure 4.1 (c) Ziehl Neelsen (ZN) stain of FNA of a tuberculous lymph node demonstrates the presence of acid fast bacilli (pink rod-like structures). ×400.

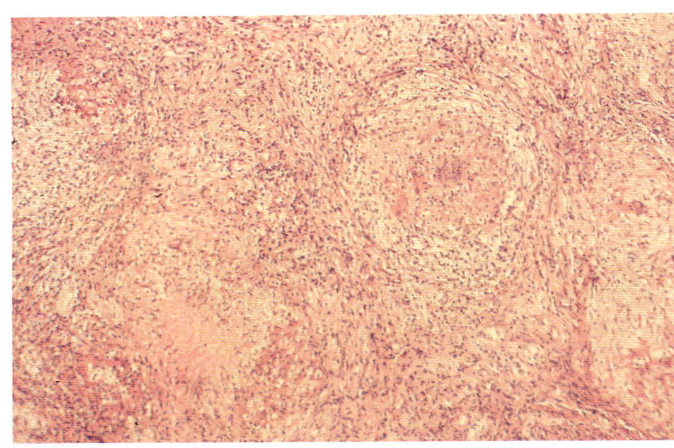

(d)

Figure 4.1 (d) Histological section of a lymph node showing granulomata of varying ages. One is well formed, with fusion of multiple elongated epithelioid cells forming a central Langhans' giant cell (LGC), surrounded by plump and slender epithelioid cells. A second granuloma shows central caseous necrosis surrounded by loosely arranged epithelioid cells. These features are those of active tuberculosis. H&E ×100.

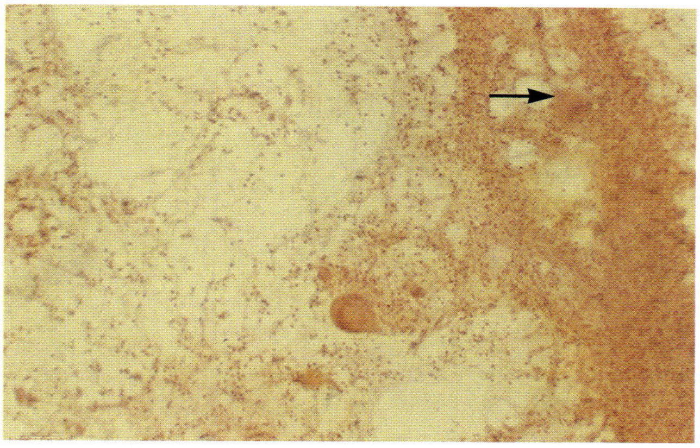

(a)

Figure 4.2 (a) FNA of a neck node showing a large LGC, with the characteristic peripheral distribution of oval epithelioid nuclei, giving it the so-called 'horseshoe' appearance. Another smaller LGC (arrow) is enmeshed in a mixture of epithelioid cells and lymphoid cells. ×100.

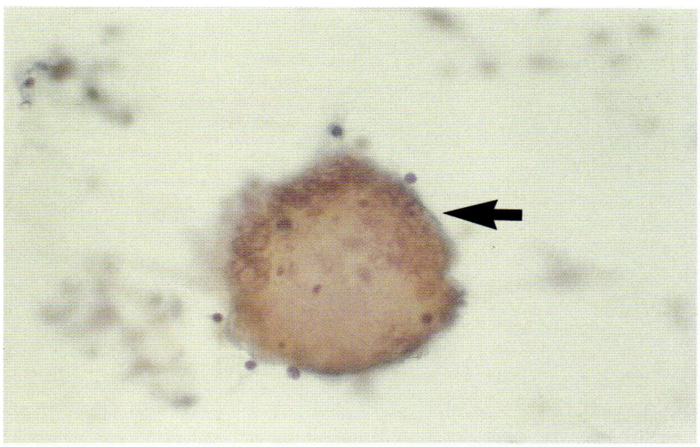

(b)

Figure 4.2 (b) High-power view of a large LGC with amorphous uniform cytoplasm and peripheral distribution of round to oval nuclei. ×400.

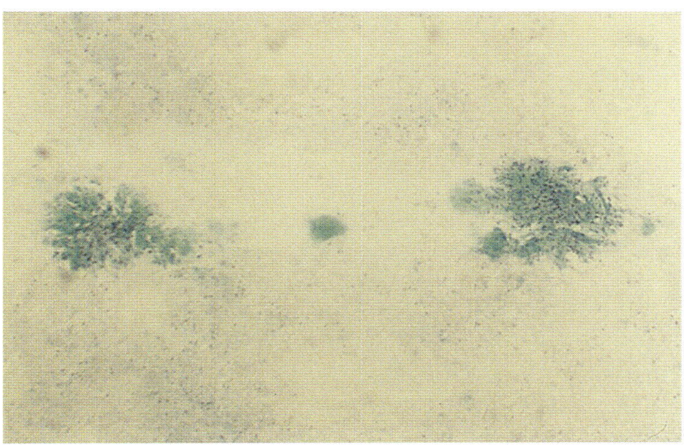

(c)

Figure 4.2 (c) FNA of an enlarged neck node in a young pregnant woman – an incidental clinical finding when the patient presented for a cervical smear. Scanning-power view of two discrete granulomatous structures separated by a clump of amorphous necrotic material. Scanty small lymphoid cells are present in the background. ×40.

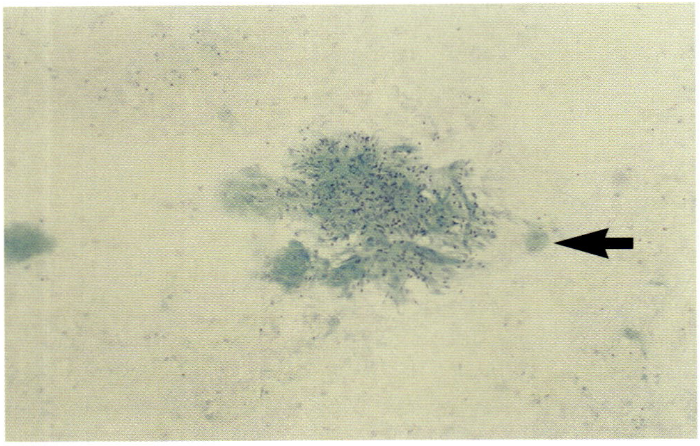

(d)

Figure 4.2 (d) Low-power view of one of the granulomata, showing the oval shape of the epithelioid cells and their nuclei. A multinucleated cell is present in the background. A sprinkling of lymphocytes is seen. ×100.

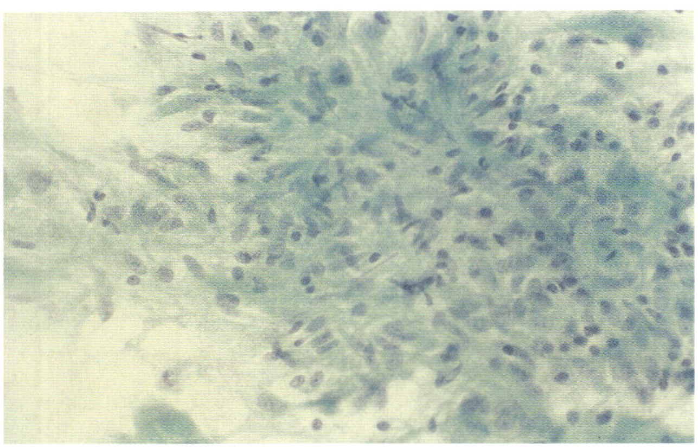

(e)

Figure 4.2 (e) High-power view of a granuloma composed of splayed-out epithelioid cells and fibroblastic cells. Scattered lymphoid cells are present. ×400.

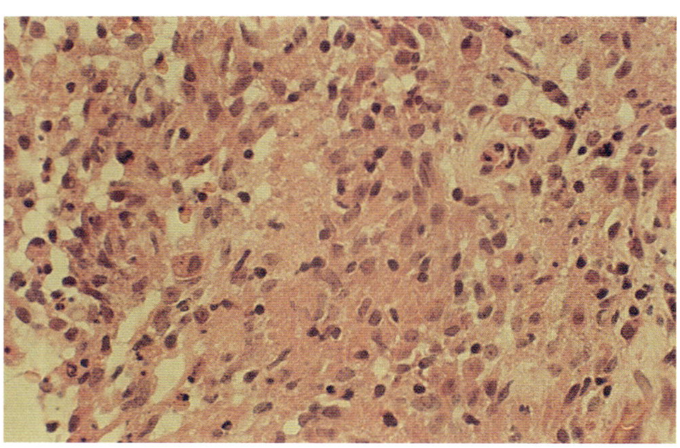

(f)

Figure 4.2 (f) Histological section of a granuloma showing similar features to those seen in (e), together with caseous necrosis. H&E ×400.

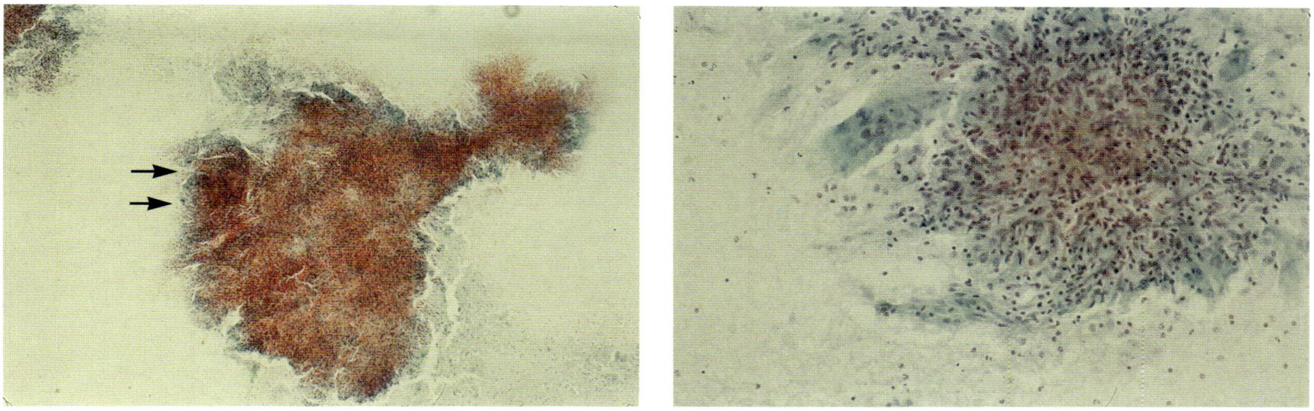

(a) (b)

Figure 4.3 (a) Scanning-power view of an FNA from an adult with lymphadenopathy in the neck. Clinical diagnosis was 'metastases from a lung mass or tuberculosis?'. This photomicrograph shows a microbiopsy of granulomatous tissue with multinucleated cells (arrows) distributed at the periphery. ×40. (b) At a higher magnification a multinucleated LGC is seen at the edge of the granulomatous tissue shown in (a). Note the splayed-out spindle-shaped epithelioid cells forming the granuloma. Small lymphoid cells are scattered about. ×400.

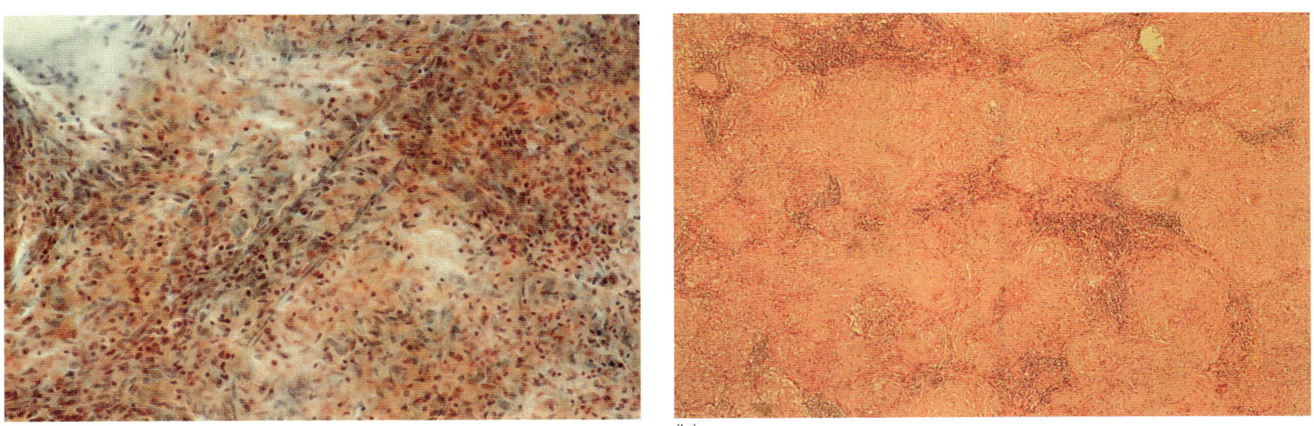

(a) (b)

Figure 4.4 (a) Another high-power view of the granulomatous tissue shown in Figure 4.3a, showing narrow capillary vessels coursing through it. ×400. (b) Histological section of a tuberculous lymph node, showing the coalescence of granulomata to form a mass of solid granulomatous tissue. H&E ×100.

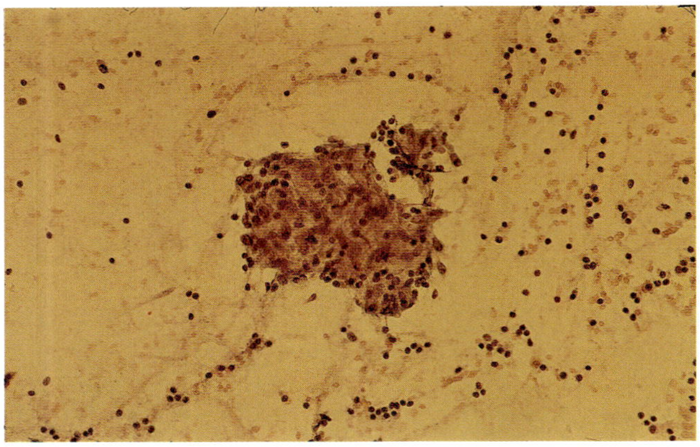

(a)

Figure 4.5 (a) FNA of a neck node in a teenager shows a discrete granuloma formed by loosely arranged, plump, round to oval epithelioid cells. Mature lymphocytes are present in the background. ×400.

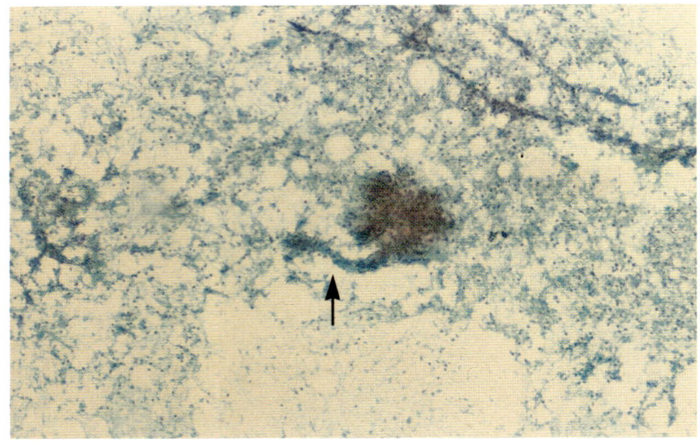

(b)

Figure 4.5 (b) Another granulomatous structure lying in a background of lymphocytes and loose necrotic material. Note the fairly thick blood vessel, which has formed to give the structure its blood supply (arrow). Granulomata often appear singly on FNA specimens. ×100.

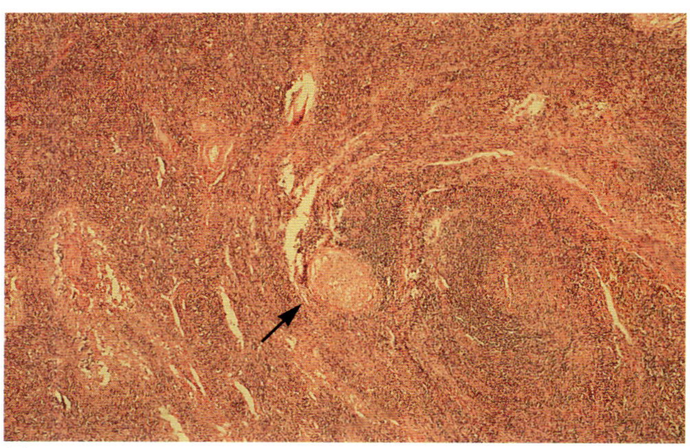

(c)

Figure 4.5 (c) Histological section of the node showing a single discrete granuloma (arrow) and indistinct granulomatous tissue within a tuberculous lymph node. It is easy to appreciate how an FNA could miss a solitary granuloma; several passes when performing the FNA ensure a representative sample. ×40.

(d)

Figure 4.5 (d) Higher magnification of the edge of the tuberculous granulomatous tissue to emphasize its indistinct delineation from the surrounding lymphoid tissue. H&E ×100.

CAT-SCRATCH DISEASE AND TOXOPLASMOSIS

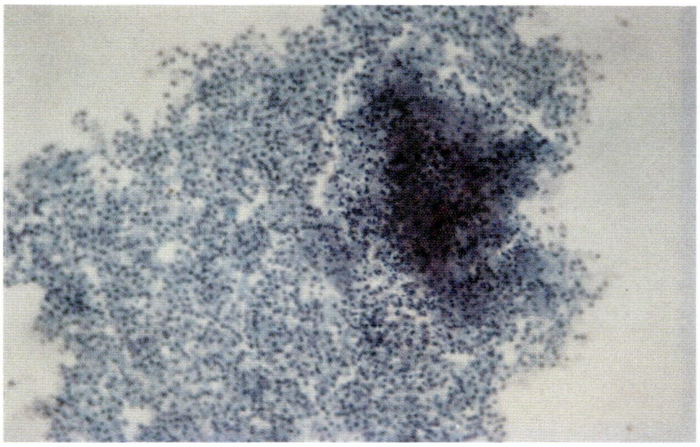

(a)

Figure 4.6 (a) FNA of a neck node from a patient who had suffered scratches on the face from a cat. The aspirate consisted almost entirely of lymphoid cells, with focal microabscess formation. ×400.

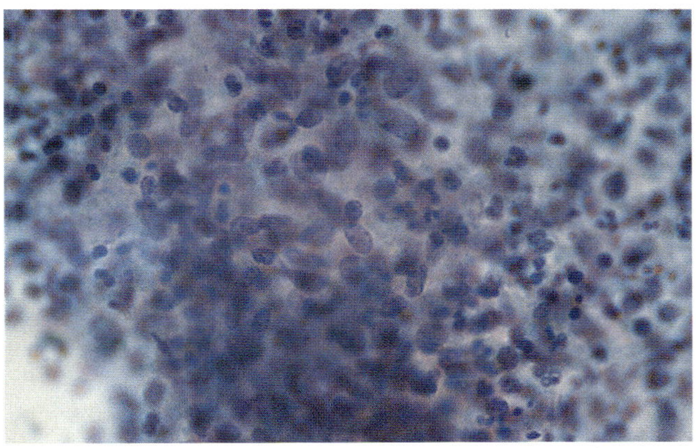

(b)

Figure 4.6 (b) High-power view of a microabscess. Note the polymorphs among the lymphocytes, which show varying maturity. No central necrosis is present. ×400.

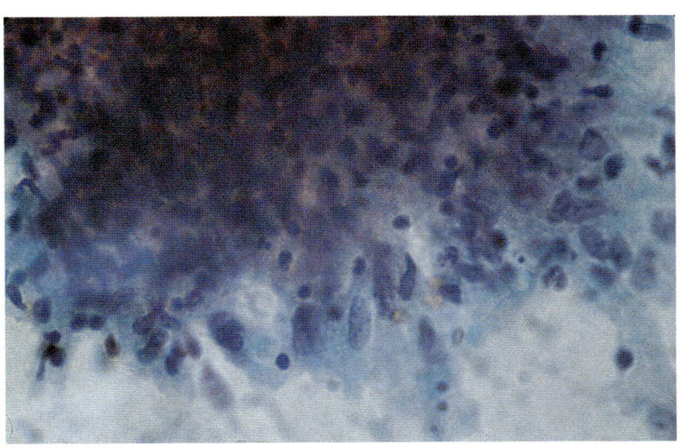

(c)

Figure 4.6 (c) At the edge of the microabscess plump fibroblasts are forming. These appearances are compatible with cat-scratch disease (Silverman, 1985). ×400.

TOXOPLASMOSIS

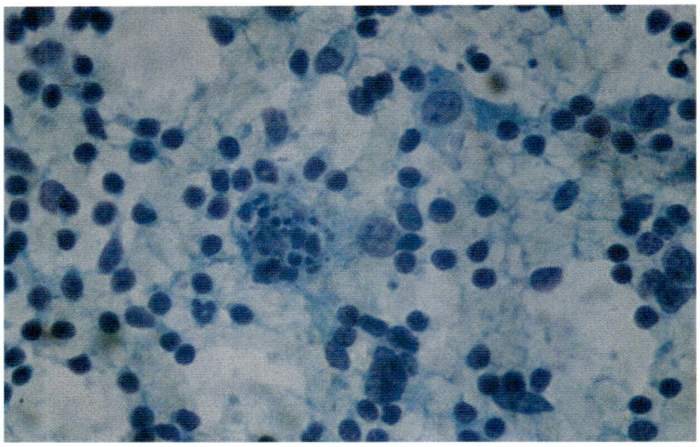

(a)

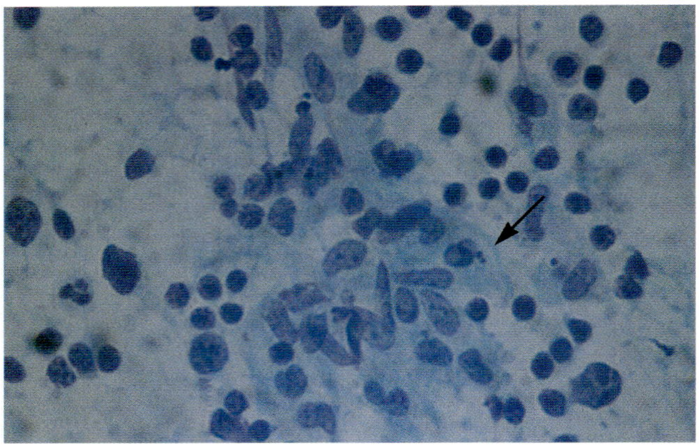

(b)

Figure 4.7 (a) and (b) FNA of a neck node in a patient with generalized lymphadenopathy. Macrophages containing cellular fragments ('tingible-body macrophages') are present among lymphoid cells. Immunoblastic cells with enlarged round nuclei; note the disturbed chromatin and prominent nucleoli, together with epithelioid cells containing elongated pale nuclei and pinpoint nucleoli. Paired intracytoplasmic bodies are seen (arrow). These appearances are compatible with toxoplasmic lymphadenitis (Chirst and Feltes-Kennedy, 1982). Serological tests for toxoplasma were positive. ×400.

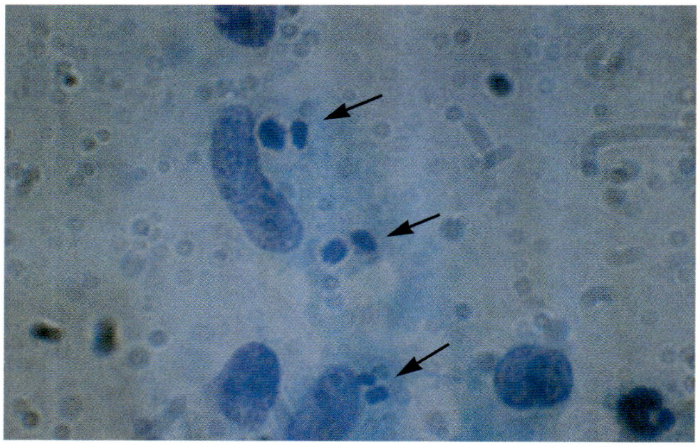

(c)

Figure 4.7 (c) High-power view of an epithelioid cell with paired intracytoplasmic toxoplasma bodies (arrows). Oil immersion. ×1000

ANTHRACOSIS OF LYMPH NODES IN HUT-DWELLERS

Severe pulmonary fibrosis associated with silica deposition resulting from chronic inhalation of smoke in hut-dwellers in rural parts of southern Africa is known as 'hut lung' (Grobbelar and Bateman, 1991). The huts have no smoke outlet: fires are lit inside the huts during the cold weather. Corn is also ground inside the huts in sandstone 'bowls'. Thus there is heavy inhalation of carbon, and silica particles. Patients of both sexes, but more commonly females, present with severe dyspnoea and are often emaciated. These patients are also found to have enlarged axillary and neck nodes (due to carbon deposition), which have clinically been suspected as being TB or metastatic carcinoma.

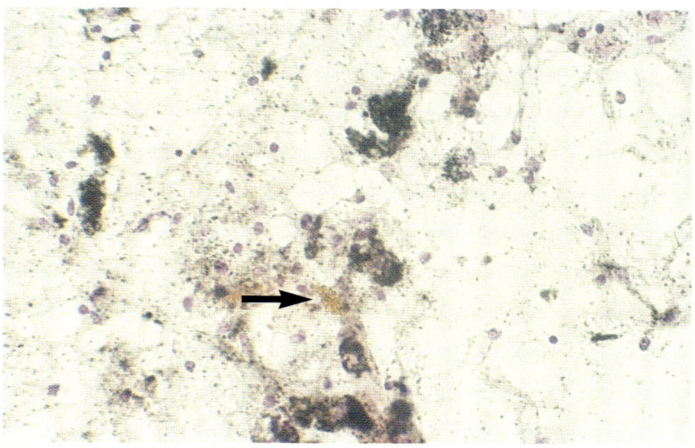

(a)

Figure 4.8 (a) FNA of a neck node from a patient with diffuse enlargement of the lymph nodes in the neck. The clinician stated that the aspirate resembled liquorice! Low-power view showed intensely black pigment. Some was loose lying and dusty, and the remainder appeared to be within macrophages. The cytotechnologist at the on-site clinic suggested a diagnosis of metastatic melanoma; however, melanin pigment is usually golden-brown/black. In the author's experience melanin is never as black as coal or carbon. Note also the granular material, possibly silica (arrow). Furthermore, the clinical background of the patient was of great importance in this case. The patient was a Xhosa hut-dweller from a rural area and had presented to the outpatient department with severe dyspnoea and emaciation. Malignant melanoma does occur in the Xhosa race, but is very rare. ×100.

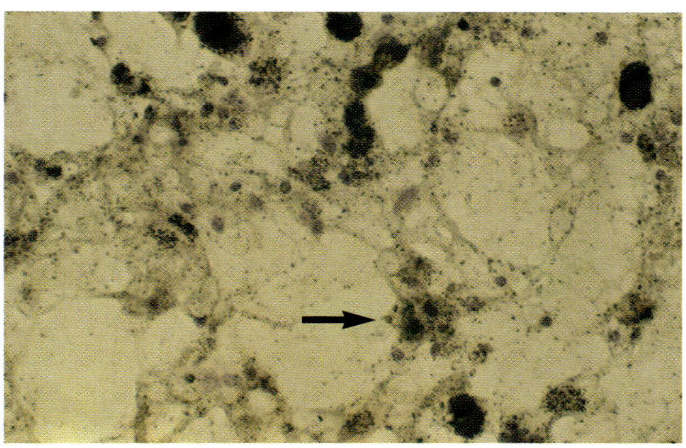

(b)

Figure 4.8 (b) High power view emphasizes the intensely black nature of the pigment. It is seen within macrophages as well as lying free. The author suggested a diagnosis of anthracosis, probably associated with 'hut lung'. Subsequent chest X-rays revealed diffuse fibrosis of the lungs compatible with this diagnosis. ×100.

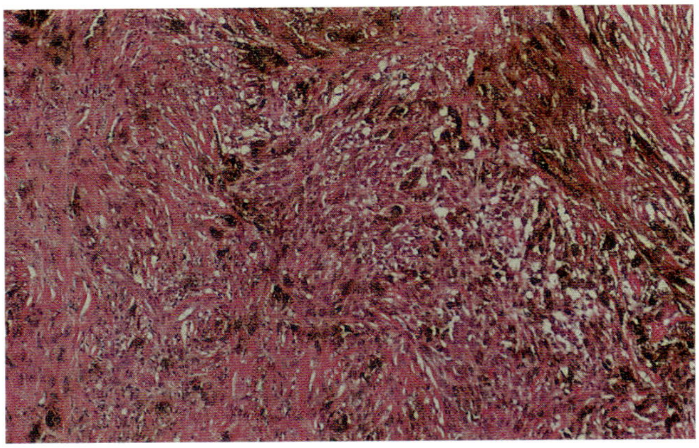

(c)

Figure 4.8 (c) Histology of the excised matted anthracotic lymph nodes revealed fragmented and sclerotic lymphoid material containing carbon pigment and scattered silicotic granules. Note dense sclerotic tissue containing carbon pigment, lymphoid tissue. The dense sclerotic reaction has been caused by the presence of carbon and silica. In other areas loose fibroblastic tissue associated with carbon and loose lying macrophages are seen. ×100.

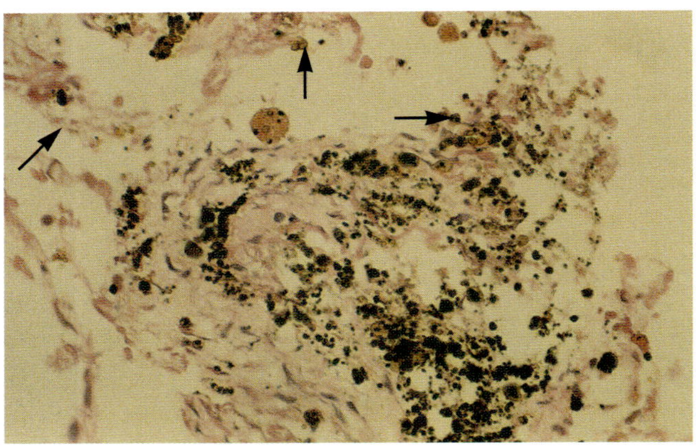

(d)

Figure 4.8 (d) High power view of the anthracotic node shown in Figures 4.8 (a) and 4.8 (b). This shows loose sclerotic tissue which is heavily infiltrated by black carbon pigment and golden yellow silica particles (arrows). ×400.

ECTOPIC THYROID TISSUE IN FNA SPECIMENS OF THE NECK

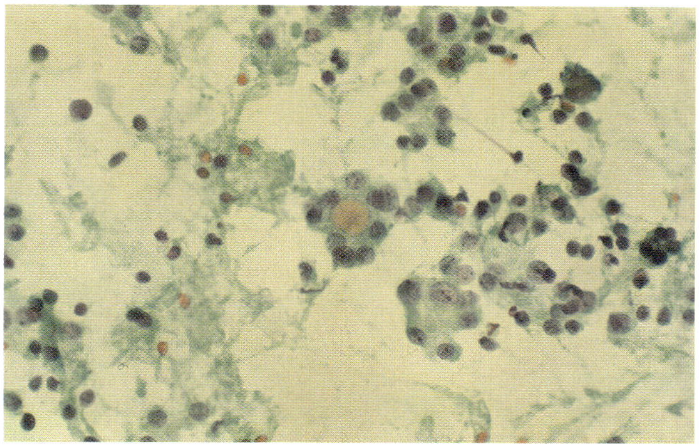

(a)

Figure 4.9 (a) FNA of a symptomless mass in the posterior triangle of the neck from a 50-year-old female. Small un form cells are present in a 'clean' background. Some of the cell groups have a follicular arrangement, with central amorphous eosinophilic material. These appearances suggest an origin from the thyroid gland or 'lateral aberrant thyroid tissue'. Remember that metastatic wel -differentiated follicular carcinoma of the thyroid may be composed of apparently normal thyroid tissue.

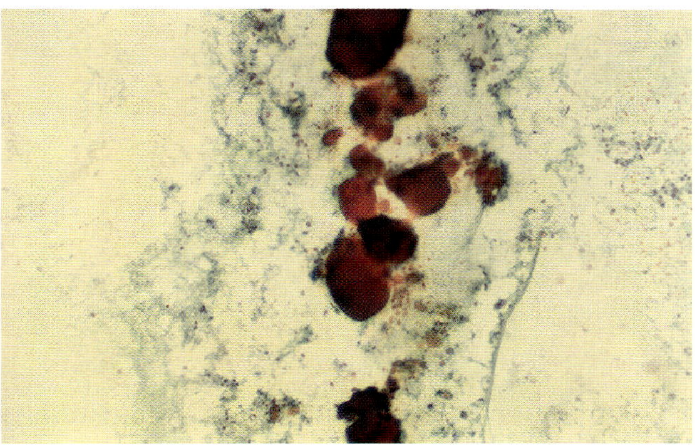

(b)

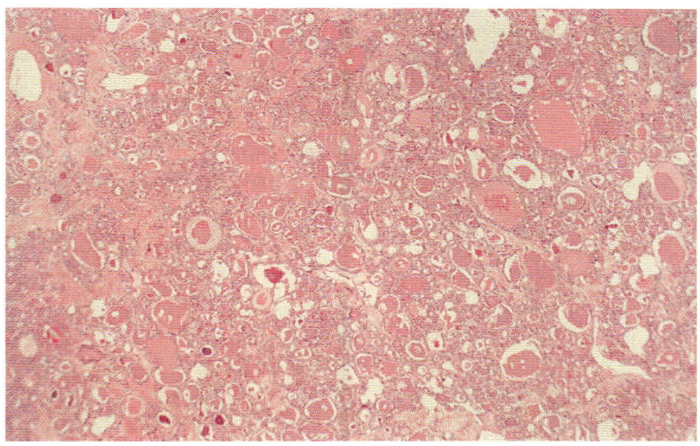

(c)

Figure 4.9 (b) and (c) Clumps of dense amorphous eosinophilic material, probably colloid, are also present. Investigation of the thyroid gland and a bone scan failed to reveal any abnormalities. Histological examination of the excised mass showed normal thyroid tissue. Continued follow-up of the patient was advised.

NECROSIS AND CRYSTALLINE MATERIAL

Crystalline material of uncertain nature is sometimes found in aspirates from necrotic neck nodes. The crystals are usually orangeophilic, elongated and hexagonal. The necrotic material does not have the features of caseous necrosis. The aetiology of the crystalline material is uncertain.

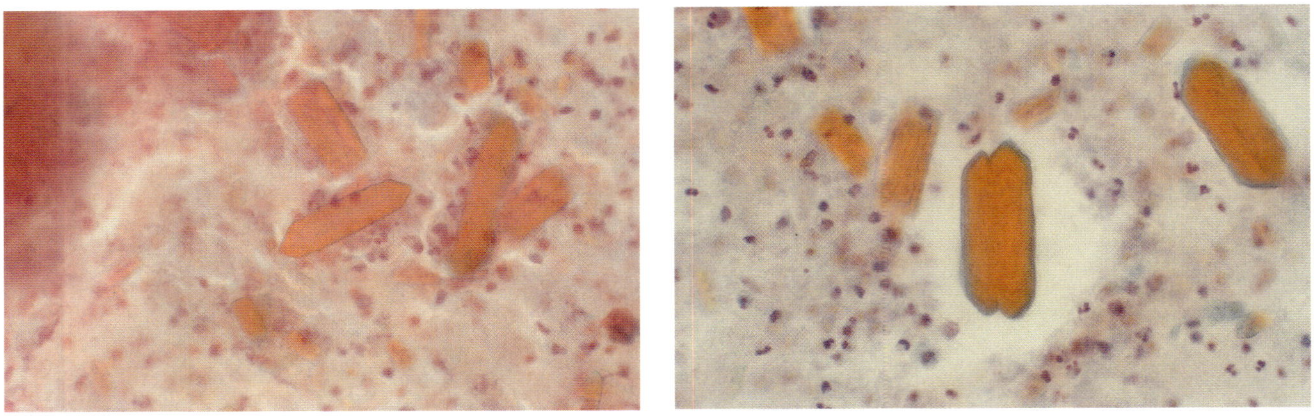

(a) (b)

Figure 4.10 (a) and (b) Large, eosinophilic, elongated, sharply angulated hexagonal crystals are present in this aspirate from a neck node in a symptomless adult male patient. ×400.

REFERENCES

Tuberculosis

Bailey, T. M, Akhtar, M. and Ali, M. A. (1985) Fine needle aspiration biopsy in the diagnosis of tuberculosis. *Acta Cytol.* **29**; 732–736.

Brown, L. A. and Coghill, S. B. (1992) FNA for the diagnosis and management of granulomatous disease. *Cytopathology* **3**: 9–15.

Finfer, M., Perchcik, A. and Burstein, D. (1991) FNA diagnosis of tuberculosis in lymphadenitis in patients with and without AIDS. *Acta Cytol.* **35**: 325–332.

Gupta, A. K, Nayar, M. and Chandra, M. (1992) Critical appraisal of fine-needle aspiration cytology in tuberculous lymphadenitis. *Acta Cytol.* **36**: 391–394.

Koss, L. (1984) *Aspiration Biopsy: Cytological Interpretation and Histologic Basis*. New York, Igaku Shoin, 109–115.

Kupper, T. H., Sreffen, U., Wehle, K., Richartz, G. and Pfitzer, P. (1995) Morphological study of bacteria of the respiratory system using fluorescence of Papanicolaou-stained smears with special regard to the identification of Mycobacteria spp. *Cytopathology* **6**: 388–402.

Shariff, S. and Thomas, J. A. (1991) FNA cytodiagnosis of clinically suspected tuberculosis in tissue enlargements. *Acta Cytol.* **35**: 333–336.

Cat-scratch disease

Silverman, J. F. (1985) Fine-needle aspiration cytology of cat-scratch disease. *Acta Cytol.* **29**: 542–547.

Toxoplasmosis

Chirst, M. L. and Feltes-Kennedy, M. (1982) Fine-needle aspiration cytology of toxoplasmic lymphadenitis. *Acta Cytol.* **26**: 425–428.

Anthracosis

Grobbelar, J. and Bateman, E. D. (1991) 'Hut lung': a domestically acquired pneumoconiosis of mixed aetiology in rural women. *Thorax* **46**: 334–340.

5 Lymph nodes – lymphoproliferative processes

INTRODUCTION

Reactive lymph node enlargement may result from a variety of antigenic stimuli; the cytological appearance of the component cells varies according to the type of immune response. FNA of a benign reactive node contains a mixture of the various types of cells which are proliferating at the time of aspiration:

- viral infections commonly cause paracortical hyperplasia
- non-specific infections cause follicular centre-cell reaction, with follicular hyperplasia and proliferation of lymphocytes of T- or B-cell type, depending on the aetiological agent
- sinus histiocytosis results from a variety of antigens
- in all three instances it is the variety of cells, i.e. small lymphocytes, immunoblasts, 'tingible-body' macrophages with intracytoplasmic nuclear debris, and plasma cells, that indicate the benign proliferative nature of the process. Follicular arrangement of lymphoid cells may be seen.

Granulomatous reactions in lymph nodes are discussed in Chapter 4. It must be stressed that the background appearances of FNA of the lymph nodes is very important. Benign proliferative nodes have a clear background. Cell debris is indicative of a cyst or malignancy. Necrosis suggests TB or malignancy (see Chapters 4 and 6).

In all cytological preparations lymphoid cells lie singly – except when an intact follicle of cells is aspirated. Lymph nodes undergoing neoplastic proliferation usually show a monotonous cell population; 'tingible-body' macrophages are virtually never seen. In large-cell anaplastic lymphoma the cell population is polymorphous, with large pleomorphic cells. They have plentiful cytoplasm and are often multinucleated. The cleaved nuclei have irregular outlines and prominent nucleoli. Differential diagnosis of large-cell lymphoma from carcinoma requires careful clinical correlation and immunocytochemical staining.

Hodgkin's disease is characterized by the presence of Reed–Sternberg (RS) cells. In its classic form the RS cell is a binucleated mirror-image giant cell with large lobulated nuclei and very large 'owl's eye' nucleoli.

Very little literature is available on the cytological diagnosis of lymphomas. In the author's experience, FNA of lymph nodes in lymphoproliferative processes has the following advantages, especially in triage management of patients:

- metastatic lesions can be excluded in the majority of cases
- selection of a representative node for surgical excision is facilitated
- it is easily repeatable, enabling assessment of response to chemotherapy
- cytological material obtained on FNA is adequate in the diagnosis of recurrent and inoperable disease
- the pathologist can be alerted prior to surgery, so that fresh imprints can be prepared for immunocytochemical staining, and the excised lymph node can be properly fixed.

In triage management it is not necessary, nor – in the author's opinion – is it possible, to give a specific diagnosis on the type of lymphoma. The diagnostic cytopathologist's duty is to assist the clinician in the triage management of patients. Excision of an intact lymph node (with intact capsule) for histological, immunocytochemical and molecular biological studies is mandatory for the primary diagnosis and subsequent clinical management of the majority of patients with suspected lymphoma.

REACTIVE FOLLICULAR HYPERPLASIA

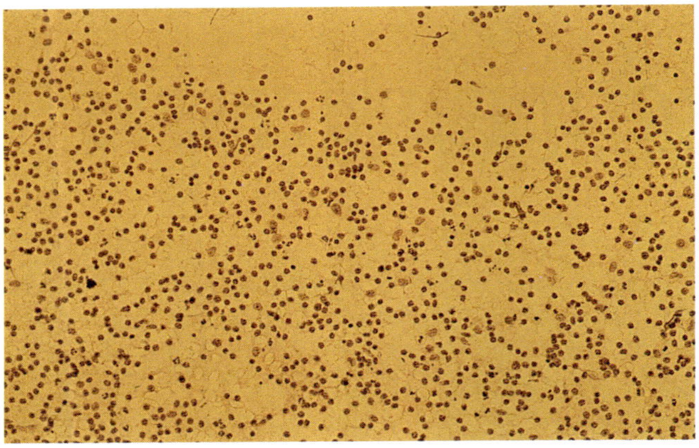

Figure 5.1 FNA of a reactive lymph node showing evenly spread, single-lying lymphoid cells in a clear background. The cells show varying degrees of maturity. The majority are small, round mature lymphocytes. Larger, pale lymphoid cells are scattered throughout. Some of these have small pinpoint nucleoli. Cleaved nuclei are not seen. There are no granulomata, giant cells or apoptotic cells. ×100.

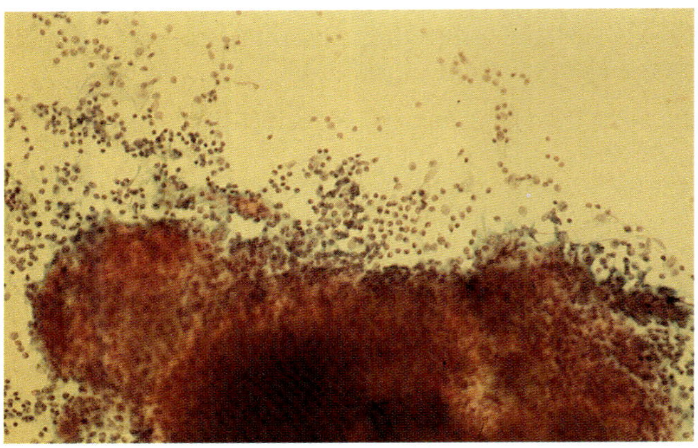

Figure 5.2 This photomicrograph shows unevenly spread lymphoid cells which, in the lower portion of the picture, show follicular arrangement. Loose, single-lying small lymphoid cells are also present, which show varying stages of maturity. These features on an FNA are compatible with benign follicular hyperplasia. The background is clear. ×100.

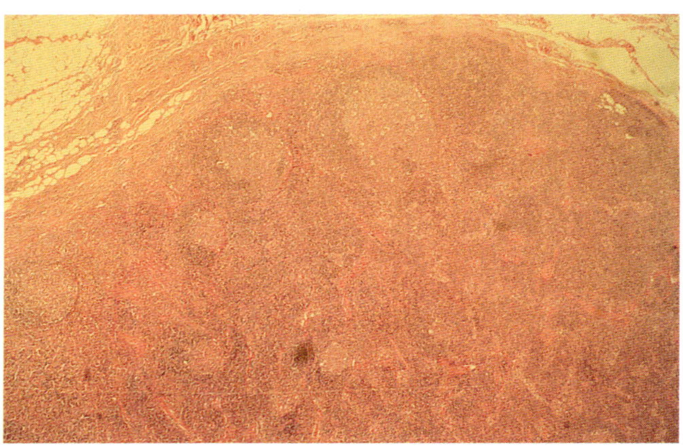

Figure 5.3 Histological section of a benign reactive lymph node showing the follicular pattern which is characteristic of follicular hyperplasia. Note the intact capsule of the node. The macrophages containing cell debris – 'tingible bodies' – appear as clear cells in the follicles. H&E ×40.

LOW-GRADE LYMPHOMA

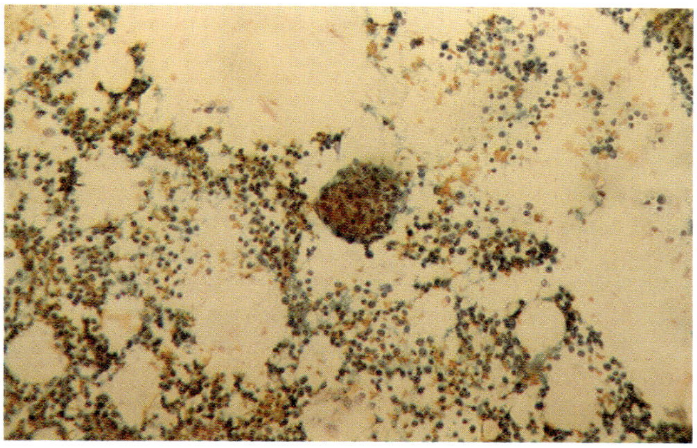

Figure 5.4 FNA of a lymph node in an elderly patient with a history of low-grade lymphoma. This confusing picture shows loose-lying, small, monotonous lymphoid cells and a three-dimensional cluster of cells, and was at first thought to be metastatic adenocarcinoma. However, careful attention to *all* the cells shows that they possess the same cytological characteristics, although some are slightly larger than others. The cluster of cells probably represents a follicle. No tingible-body macrophages, debris or necrosis are present. In view of the clinical history, a low-grade lymphoma is the most likely diagnosis. ×100.

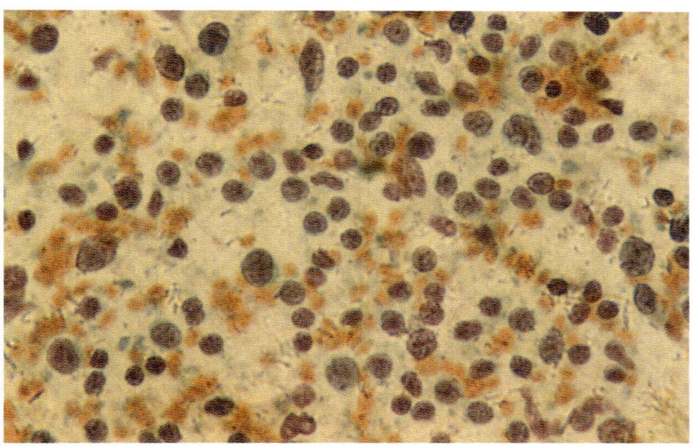

Figure 5.5 High-power view of the FNA shown in Figure 5.4. These cells are single lying and all have the same cytological characteristics, apart from slight variability in size. ×400.

HIGH-GRADE LYMPHOMA

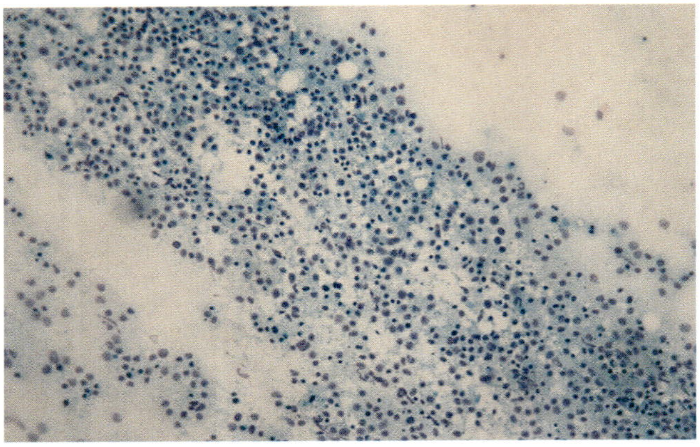

Figure 5.6 FNA of a postauricular node in a 60-year-old woman with a previous history of carcinoma of the uterine cavity and a carcinoma of the rectum. She presented with otitis media and was found to have an enlarged node, which was aspirated. This showed innumerable large, single-lying lymphoid cells, some showing degenerative changes (enlarged and pale) and others showing apoptosis. No cell clusters or necrosis are seen. These appearances are more suggestive of a high-grade lymphoma, rather than a metastatic adenocarcinoma. ×100.

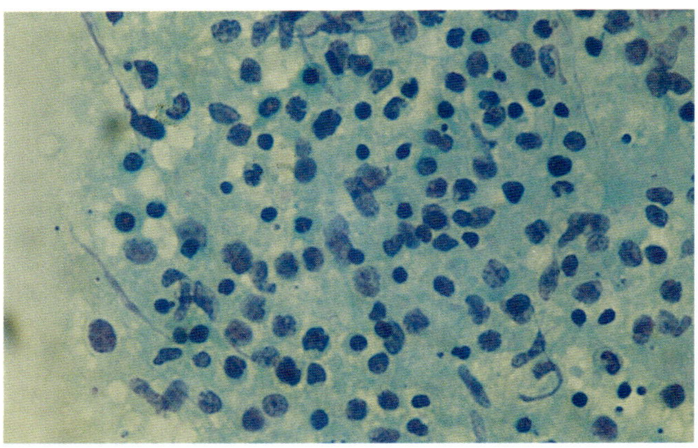

Figure 5.7 High-power view of the FNA discussed in Figure 5.6, showing the single-lying lymphoid cells, some with irregular nuclear outlines. The paler degenerate cells show 'unfolding' of cleaved nuclei to form elongated shapes. This phenomenon is not seen in histological sections as the cells are not intact after sectioning. Small, dark, apoptotic nuclei are also seen. These features are compatible with a high-grade lymphoma. ×400.

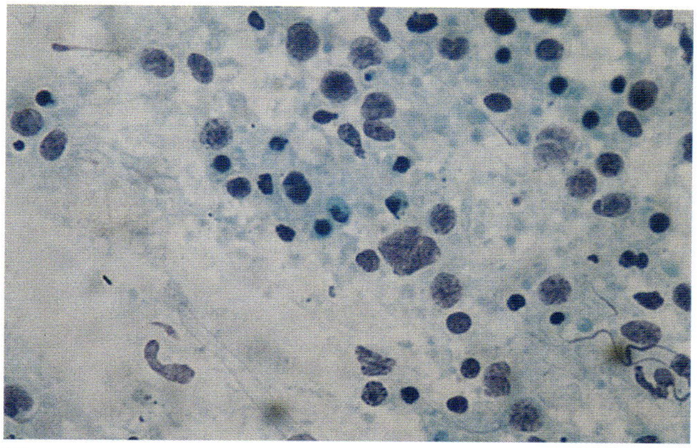

(a)

Figure 5.8 (a) High power view of the FNA discussed in 5.6 shows the single lying lymphoid cells, some with irregular nuclear outlines. The paler degenerate cells show 'unfolding' of cleaved nuclei to form elongated shapes. This phenomenon is not seen in histological sections as the cells are not intact after sectioning. Small dark apoptotic nuclei are also seen. These features are compatible with a high grade lymphoma. ×400.

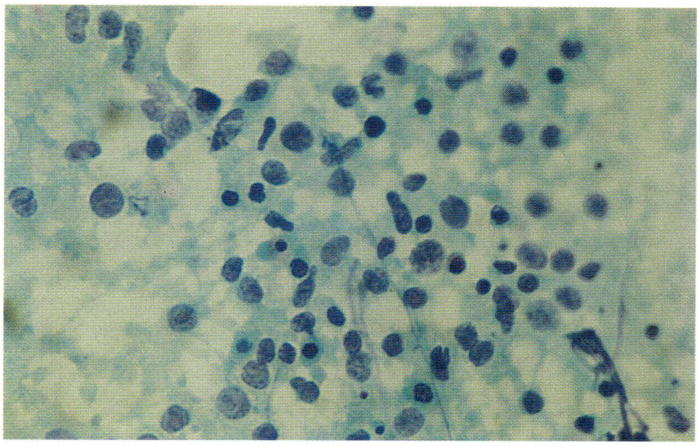

(b)

Figure 5.8 (b) Another view of an FNA from a high grade lymphoma. Large single lying nuclei, with irregular outlines, disturbed chromatin and small nucleoli. Unfolded nuclei and apoptotic nuclei are also seen, indicating a high grade lesion. ×400.

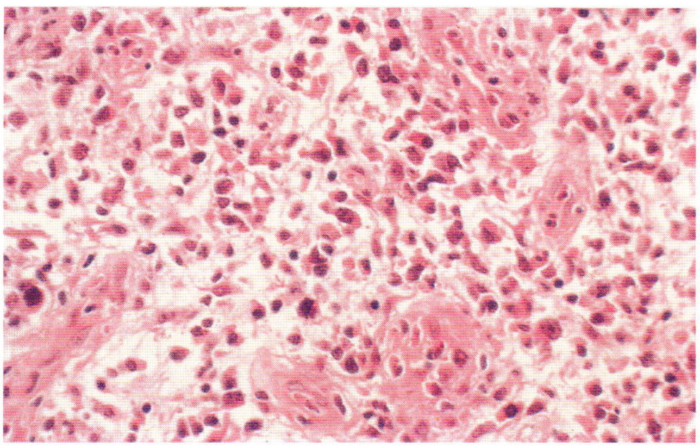

Figure 5.9 Histological section of the enlarged preauricular node discussed in Figures 5.6–5.8 showed effacement of nodal architecture by a large-cell lymphoma. H&E ×100.

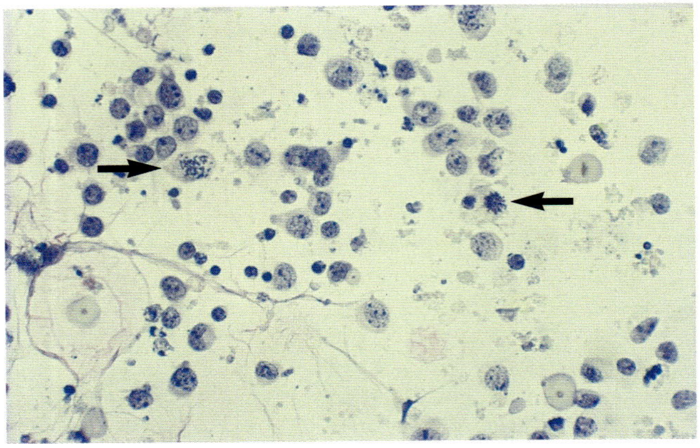

Figure 5.10 FNA of a high-grade lymphoma in a young adult. The single-lying large cells have large, round, pale nuclei, disturbed nuclear chromatin and prominent multiple nucleoli. Very striking abnormal mitotic figures are shown (arrows). There are apoptotic nuclei, 'ghost cells' and cellular debris lying in the background. None of the nuclei appear to be cleaved. ×400.

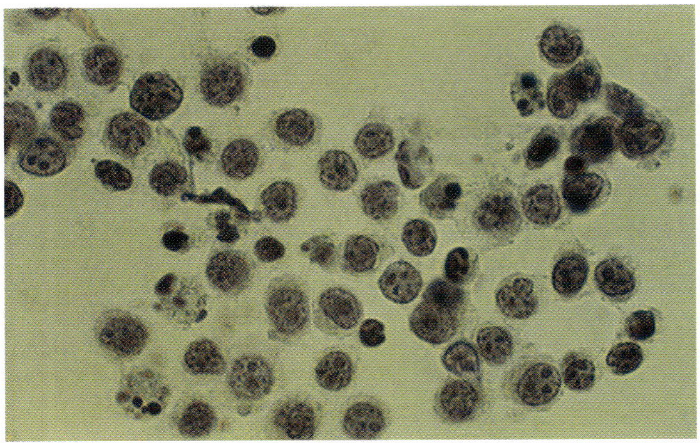

Figure 5.11 High-power view of a high-grade lymphoma showing large, single-lying, round cells with a rim of cytoplasm. The nuclear outlines are regular. Note the disturbed nuclear chromatin, the peripheral condensation of nuclear chromatin and the large, sometimes multiple, nucleoli. Apoptosis is present. These features suggest a differential diagnosis of large-cell lymphoma, metastatic melanoma, metastatic large-cell undifferentiated carcinoma, and even a variant of Hodgkin's lymphoma. Careful clinicopathological correlation is advised in a case such as this. ×630.

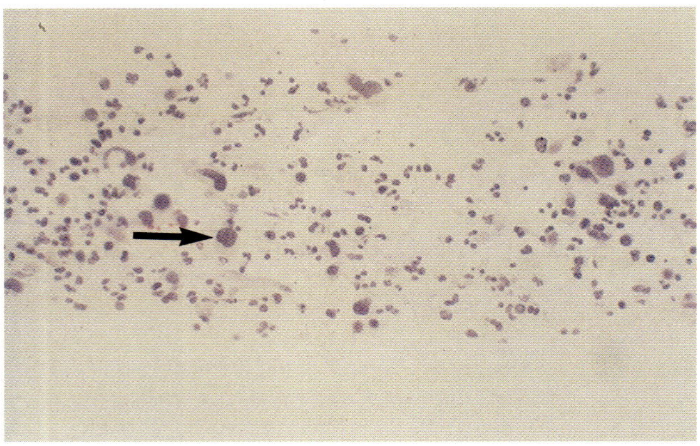

Figure 5.12 Intraoperative smear of an air-dried Giemsa-stained smear of a mixed small- and large-cell lymphoma. Note that the nuclear details are obscured and that interpretation lies solely on the arrangement of the cells (in this case single lying) and their shape and size. No comment can be made on the quality or distribution of the nuclear chromatin, or on the presence or absence of nucleoli. Giemsa. ×100.

HODGKIN'S LYMPHOMA

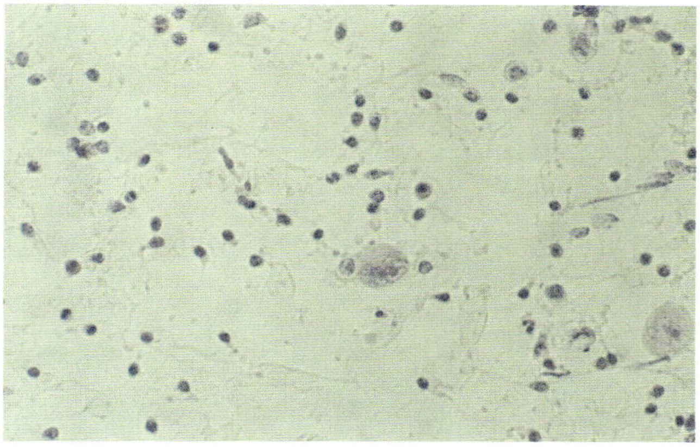

Figure 5.13 FNA of a mass in the parotid gland, which yielded very scanty cellular material, shows scattered small uniform lymphoid cells, two apparent eosinophils and a single large pale cell with pale nuclear chromatin and prominent nucleoli. This is reminiscent of a Reed-Sternberg cell. The background is clear and there are no apoptotic cells. ×400.

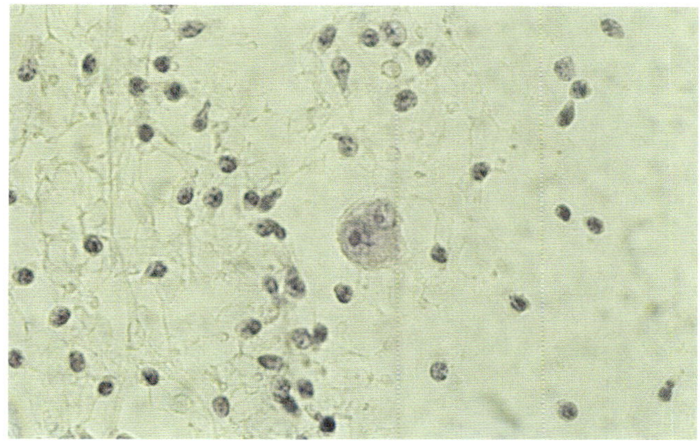

Figure 5.14 Another view of the same specimen from the parotid gland shown in Figure 5.13, showing a s ngle characteristic RS cell with an 'owl's-eye' appearance. The two large pale nuclei are lying close together in an eccentric position in the cell. There is a prominent peripheral rim of chromatin and the remainder of the nucleus is pale. The large nucleoli have an eosinophilic tinge. These are unequal in size and shape. All these features are compatible with Hodgkin's lymphoma, probably occurring in an intraparotid lymph node. Histological assessment is obviously essential in this unusual case. ×400.

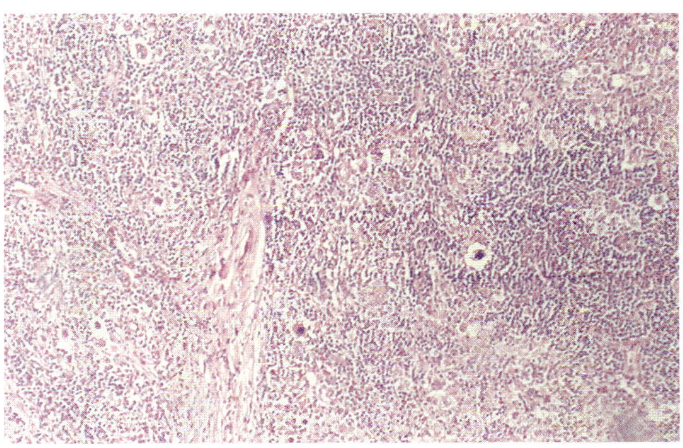

(a)

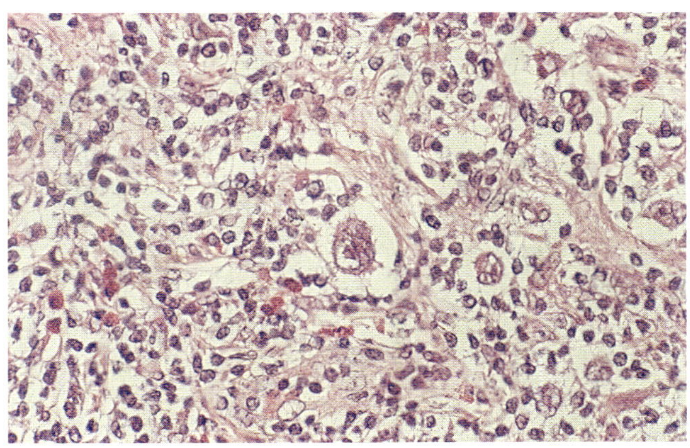

(b)

Figure 5.15 (a) and (b) Histological section of the excised parotid mass showed typical RS cells lying in sheets of lymphoid cells, and occasional eosinophils, compatible with Hodgkin's disease in intraparotid lymphoid tissue. Bands of sclerotic tissue are separating the cells, suggesting sclerosing-type Hodgkin's disease. This could account for the scanty cellular material yielded on the FNA. ×40 and ×100.

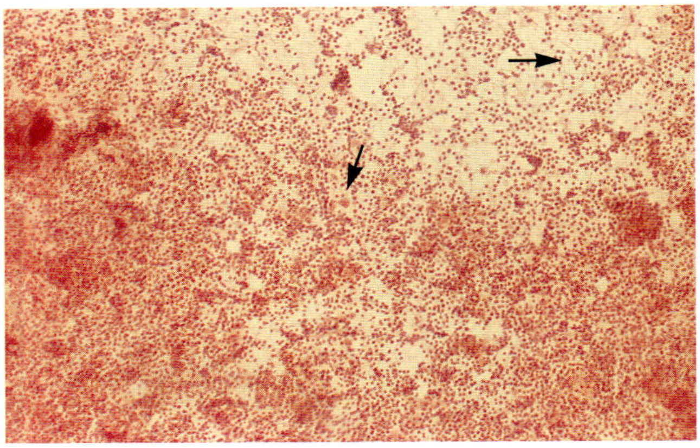

Figure 5.16 FNA of a neck node in a patient with generalized lymphadenopathy. This shows a very cellular aspirate (in contrast to Figures 5.13 and 5.14). At low power all the cells are small round lymphoid cells, with occasional isolated larger pale cells (arrows). Three-dimensional clusters of cells are also noted. However, they do not have the appearance of epithelial cells. They may represent follicular clusters of lymphoid cells. H&E ×100.

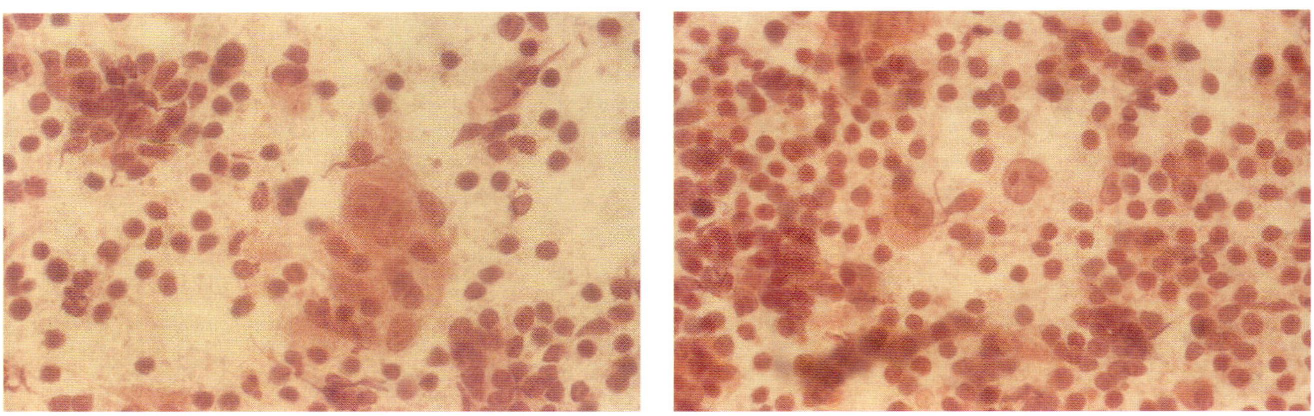

(a) (b)

Figure 5.17 (a) and (b) High-power view of the case shown in Figure 5.16 shows a group of large pale cells with large pale nuclei, peripheral condensation of the chromatin and large eosinophilic nucleoli. These cells resemble RS cells. One cell is binucleate with an 'owl's-eye' appearance. These features are suggestive of Hodgkin's lymphoma or a non-Hodgkin's lymphoma L and H type. Excision of the node and triage to the lymphoma clinic was advised. H&E ×400.

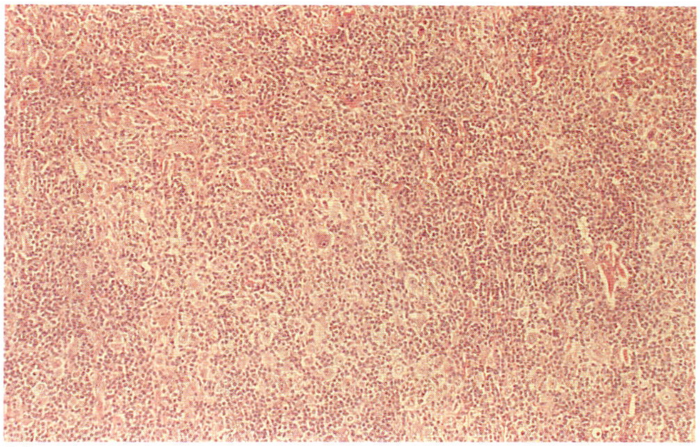

(a)

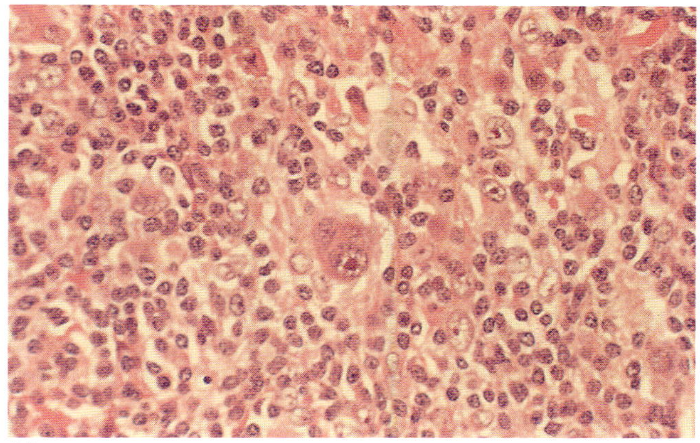

(b)

Figure 5.18 (a) and (b) Histological sections of the case shown in Figures 5.16 and 5.17 showing effacement of the lymph node by small, uniform lymphoid cells and a mixture of scattered large, pale histiocytic-type cells and RS cells. Sclerosis is not a feature (in contrast to the case illustrated in Figures 5.13 and 5.14). This probably accounts for the marked cellularity of the aspirate. H&E ×100 and ×400.

FURTHER READING

Arendse, M. P. (1993) The role of FNAB in the management of Lymphadenopathy. MMed Thesis, University of Cape Town.

Carter, T. R., Feldman, P. S., Innes, D. J., Frierson, H.F. and Frigy, A.F. (1988) The role of fine-needle aspiration in the diagnosis of lymphoma. *Acta Cytol.* **32**: 848–853.

Layfield, L. J. and Glasgow, B. J. (1991) Diagnosis of salivary gland tumours by fine needle aspiration cytology: a review of clinical utility and pitfalls. *Diagn. Cytopathol.* **7**: 267–272.

Noordyn, L. A., van Heerde, P. and Meyer, C. J. L. M. (1990) Cytology of Ki-1- (CD 30) positive large-cell lymphoma. *Cytopathology* **1**: 297–304.

Skoog, L. and Tani, E. (1991) The role of fine-needle aspiration in the diagnosis of non-Hodgkin's lymphoma. *Diagn. Oncol.* **1**: 12–18.

Sneige, N., Dekmezian, R. H., Katz, R. L., Fanning, T. V., Lukeman, J. L., Ordonez, N.F. and Cabanilles, F. F. (1990) Morphologic and immunocytochemical evaluation of 220 fine-needle aspirates of malignant lymphoma and lymphoid hyperplasia. *Acta Cytol.* **34**: 311–322.

Steel, B. L., Schwartz, M. R. and Ramzy, I. (1995) Fine-needle aspiration biopsy in the diagnosis of lymphadenopathy in 1103 patients. Role, limitations and analysis of diagnostic pitfalls. *Acta Cytol.* **39**: 76–81.

6 *Oral cavity*

INTRODUCTION

Superficial lesions of the oral cavity are readily sampled with cytological brushes; intraoral solid and cystic masses may be aspirated using traditional FNA techniques.

The surface epithelium of the oral cavity consists largely of stratified, non-keratinizing squamous epithelium. The submucosa contains varying numbers of salivary glands, which secrete a mixture of serous and mucous fluid. This generally bathes the surface squamous cells, thereby clearing away most of the inflammatory and necrotic debris which often obscures cells in exfoliative cytology of the uterine cervix. Skeletal muscle fibres are found in the deeper layers of the submucosa. The squamous cells lining the cheeks are often rich in glycogen. Areas of keratinization may be found here, associated with ill-fitting dentures or habitual cheek-biting.

The hard palate is covered by keratinizing squamous epithelium, whereas the oral surface of the soft palate is non-keratinizing. The latter gradually undergoes a transition to ciliated columnar epithelium at the posterior edge. The tongue is covered by keratinizing squamous epithelium.

COLLECTION OF SPECIMENS

Cytobrushings are obtained by gently brushing the lesion with a straight-handled cytobrush (Figure 6.1a) and then rolling the brush on to a glass slide. Routine Papanicolaou stain is used in all cases; in rare, interesting or unusual cases the slides can be destained for special stains. The specimens are fixed immediately with alcohol spray fixative. No special slide preparation was done on the cases illustrated as most of the cytobrushings were performed in a very busy clinic (in an effort to triage patients, reduce waiting time for diagnosis and treat patients on an outpatient basis, thereby providing a cheaper, more efficient diagnostic service). The cytobrush can be rinsed in a suitable fluid for bacteriological studies.

Oral cytology has been reported by several authors (Silverman *et al.*, 1977; Erenmemisoglo *et al.*, 1995; Reddy *et al.*, 1977); however, its use has limited popularity and so the total number of cases studied in each trial has been low.

The overall diagnostic rate is improved by the combined use of clinical information, cytology and histology (CICH). Definitive treatment should never be undertaken without correlation of all three diagnostic modalities (Learmonth *et al.*, 1990).

(a)

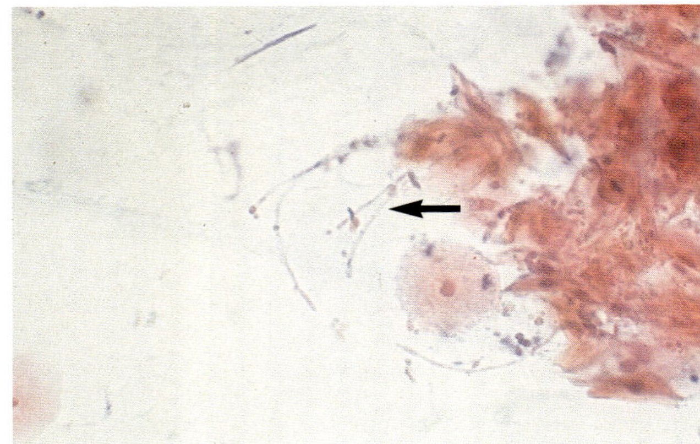

(b)

Figure 6.1 (a) Straight handled cytobrush. (b) Mild squamous atypia associated with *Candida* hyphae from a patient with oral moniliasis. The presence of hyphae is indicative of pathogenicity, whereas the presence of candida-like yeasts alone in the oral cavity is not diagnostic of infection. The hyphae illustrated show a 'nibbled' appearance owing to their digestion by saprophytic organisms. This cytological picture has been likened to 'corn on the cob'. ×100.

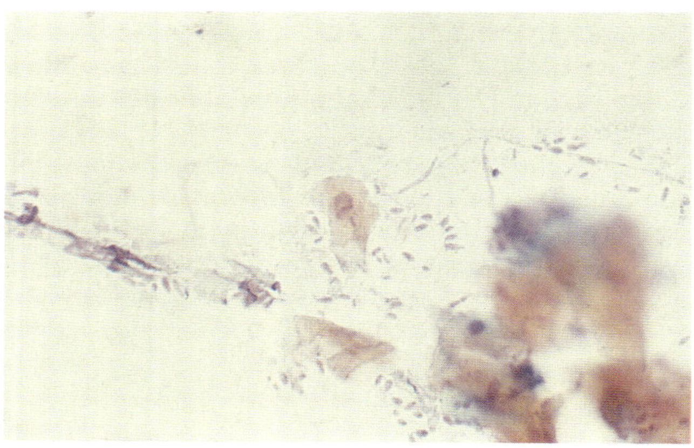

Figure 6.2 Mild squamous atypia associated with *Candida* hyphae and spores (see also Figure 7.16, page 97). ×100.

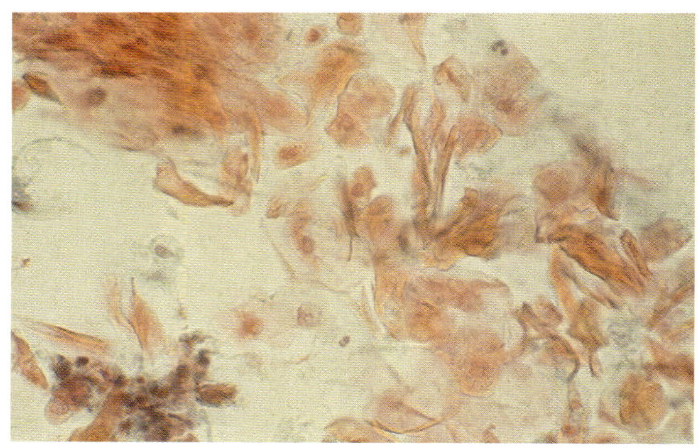

Figure 6.3 Keratinous flakes and superficial atypical orangeophilic keratinized cells from an area of chronic irritation in the mouth. ×100.

SQUAMOUS DYSPLASIA

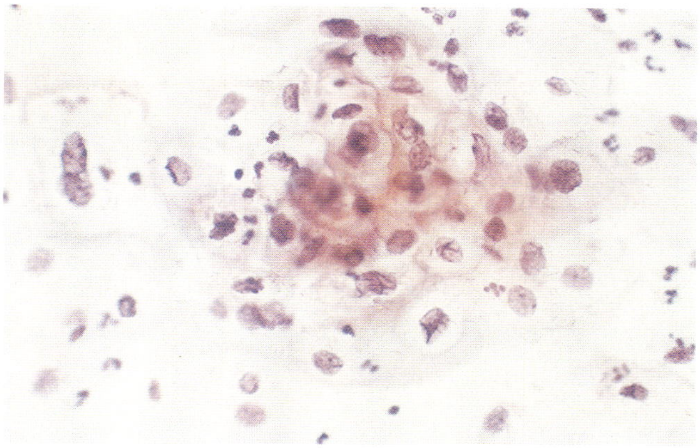

Figure 6.4 Cytobrushings from an erosive lesion in the mouth. A group of dysplastic cells, many showing koilocytic changes: plentiful cytoplasm, well-defined perinuclear haloes, enlarged pleomorphic nuclei with wrinkled nuclear membranes, and a few binucleate cells. These features are similar to those seen in human papilloma virus-associated atypia in cervical smears. Nucleoli are prominent but small and pinpoint in nature. These cells are from the case illustrated in Figure 6.5. ×100.

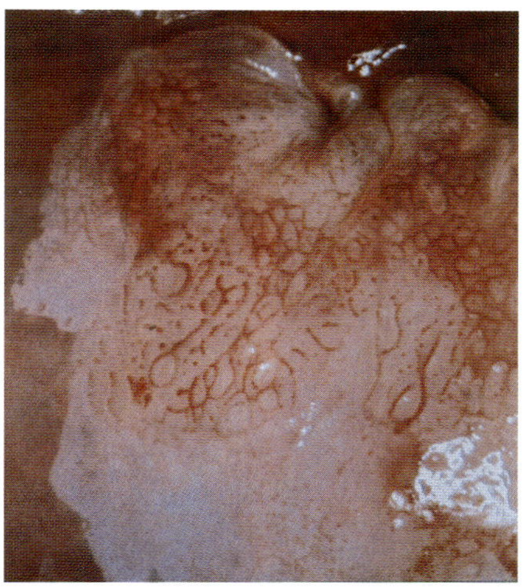

Figure 6.5 Clinical photograph of erythroplasia of the tongue. The mucosa is white and thickened. The reddened punctate appearance is ascribed to the presence of very superficial capillary blood vessels lying directly underneath, and in close apposition to, the basement membrane of severely dysplastic squamous epithelium. Its appearance is similar to that seen in foci of severe dysplasia on the surface of the uterine cervix.

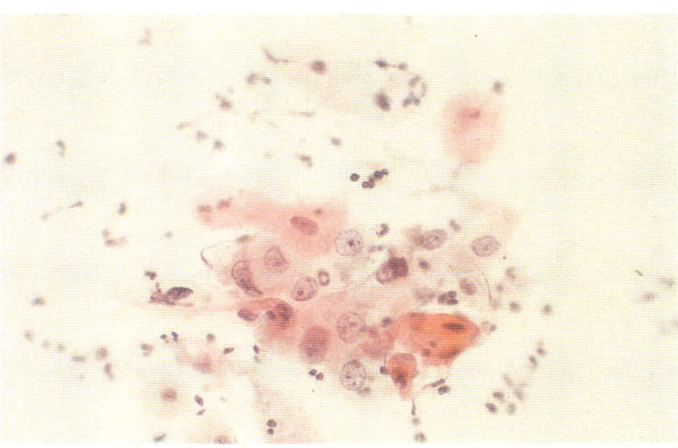

Figure 6.6 Severe dysplasia with possible invasion. Highly atypical squamous cells: pleomorphic in size and shape, loose-lying, altered N/C ratio, disturbed chromatin with large prominent nucleoli. In cytology of the oral cavity assessment of background is not as important as it is in cervical cytology, as the inflammatory/necrotic debris is often washed away by saliva. However, as in all areas of cytological interpretation, the background should always be noted. ×100.

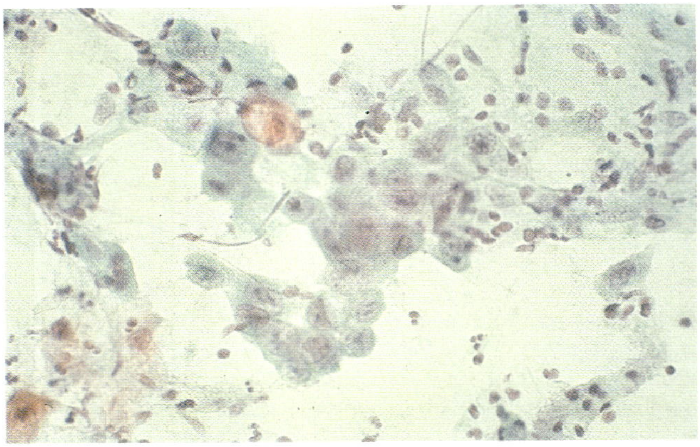

Figure 6.7 Cytobrushing of a buccal ulcer. Dysplastic cells lying in an inflammatory background with mucoid streaks. The cells are fairly cohesive, with rounded outlines, enlarged pleomorphic nuclei and prominent nucleoli. Some nuclear overlapping is present, suggestive of a severe dysplasia/carcinoma in situ. Occasional eosinophilic, dysplastic keratinized cells are present. These features raise suspicion of invasion. Careful histological assessment is advised in cases such as these, in order to plan appropriate treatment. ×100.

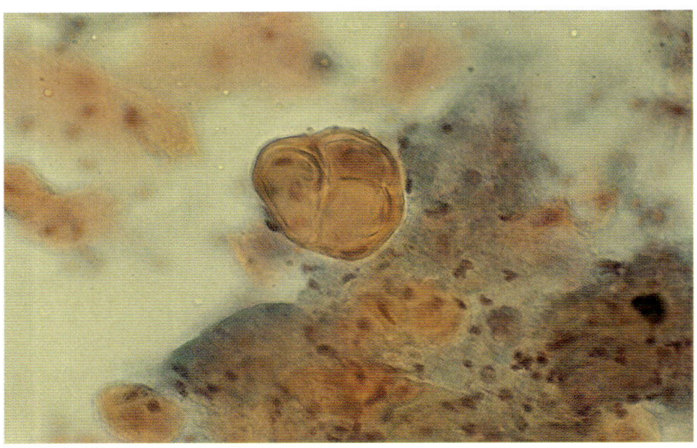

(a)

Figure 6.8 (a) A squamous pearl seen with a background of dysplastic squamous cells. ×100.

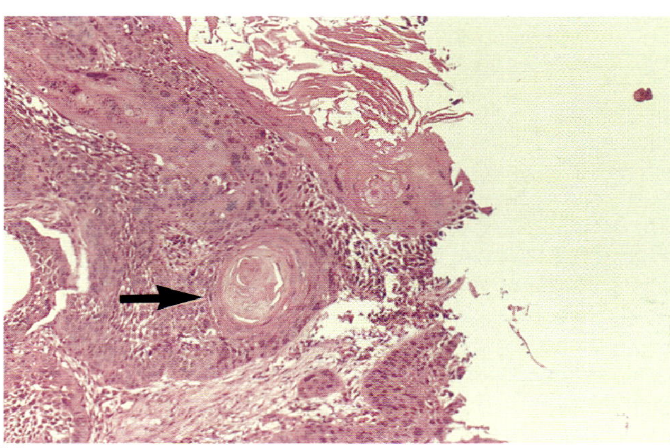

(b)

Figure 6.8 (b) Histological section showing a squamous pearl within severely dysplastic epithelium. ×100.

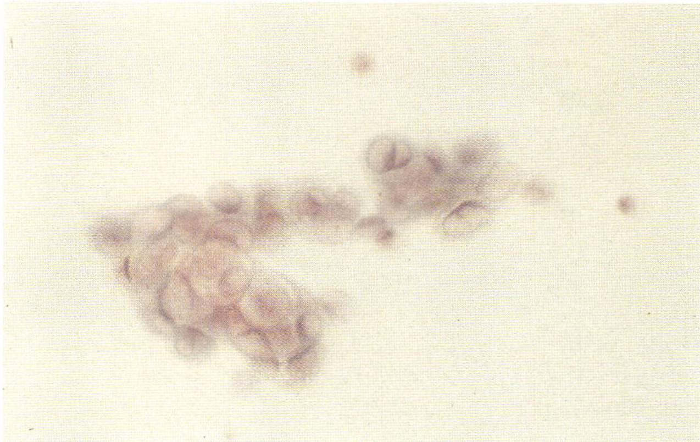

Figure 6.9 A group of small dysplastic cells showing cell-in-cell arrangement and overlapping of nuclei. Note the viral appearance: pale, 'ground-glass' appearance of the nuclear chromatin, peripheral condensation of the latter on the nuclear rim, binucleate cells. ×100.

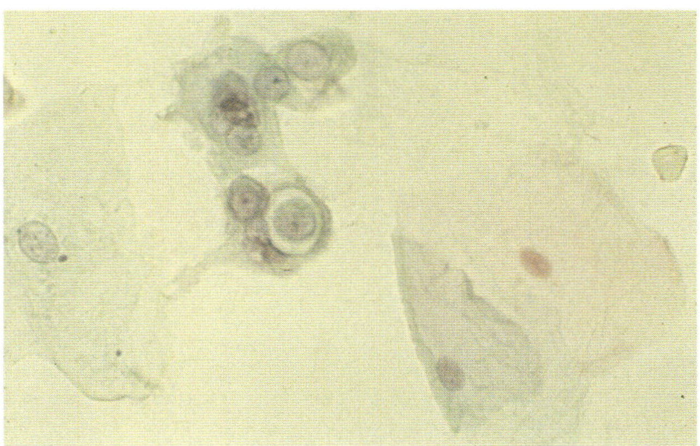

Figure 6.10 A small group of dysplastic cells with enlarged nuclei, altered chromatin, altered N/C ratio and crowding of nuclei. Note cell-in-cell arrangement. Adjacent normal squamous cells can be used for comparison. ×100.

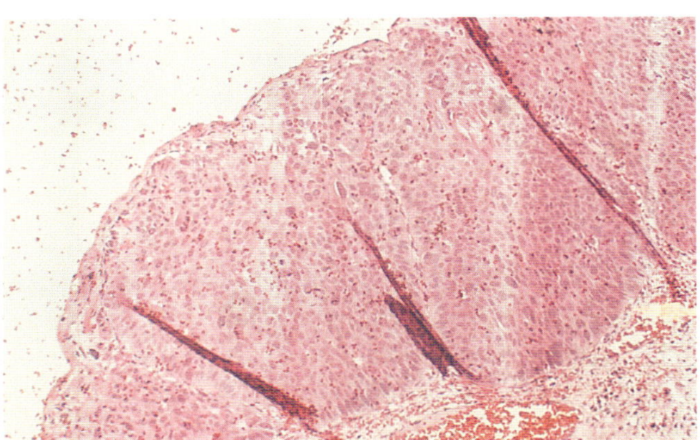

Figure 6.11 Histological appearance of the lesion depicted in Figures 6.9 and 6.10. There is full-thickness dysplasia: the cells are arranged in a disorderly fashion – 'architectural anarchy' – there is no attempt at squamous cel maturation in the upper layers of cells, the nuclei are enlarged and pleomorphic, and there are congested small blood vessels lying in close apposition to the basement membrane. The features are those of severe dysplasia. The basement membrane is intact. H&E ×100.

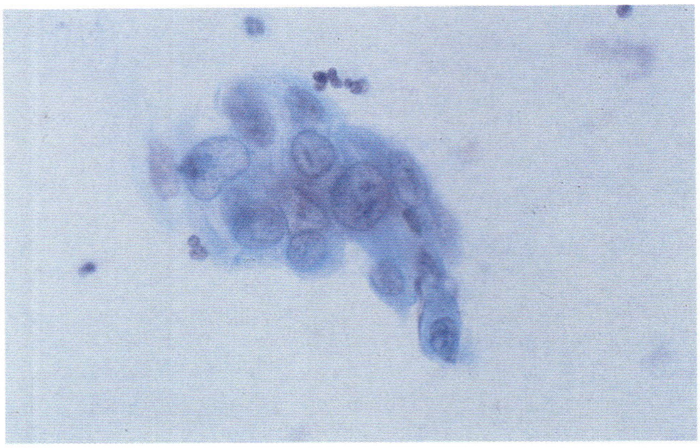

Figure 6.12 A crowded group of pleomorphic dysplastic cells with a high N/C ratio. Although the nuclei are overlapping, there are no obvious nucleoli. A high-grade lesion was diagnosed. ×100.

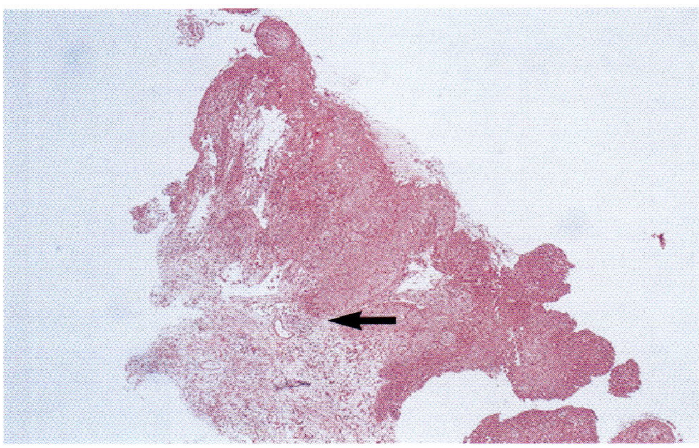

Figure 6.13 Histology section of the lesion portrayed in Figure 6.12. Sections show a fragmented biopsy with severe dysplasia. Assessment for microinvasion is difficult. Note the superficial blood vessels in the stroma. H&E ×40.

VERRUCOUS CARCINOMA IN THE ORAL CAVITY

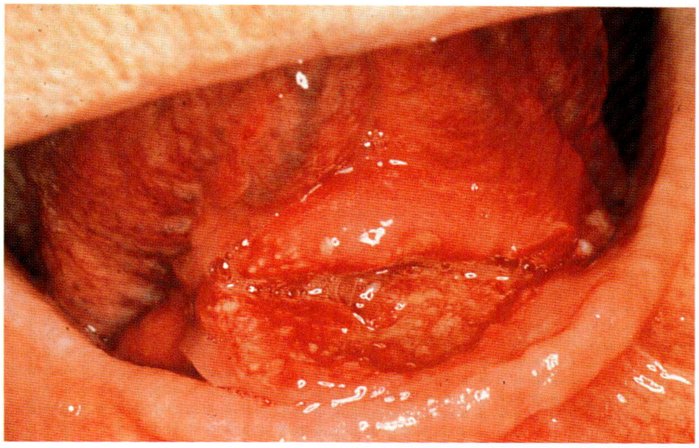

Figure 6.14 A clinical photograph of a verrucous carcinoma in the buccal sulcus.

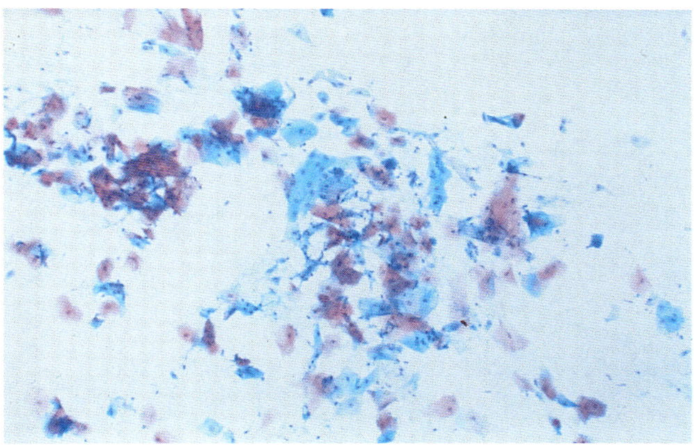

Figure 6.15 Cytobrushing of the lesion seen in Figure 6.14. Cytological atypia is minimal in these superficial squamous cells. This phenomenon of mild squamous atypia in exfoliative cytology of verrucous carcinomas of the cervix is well documented. ×100.

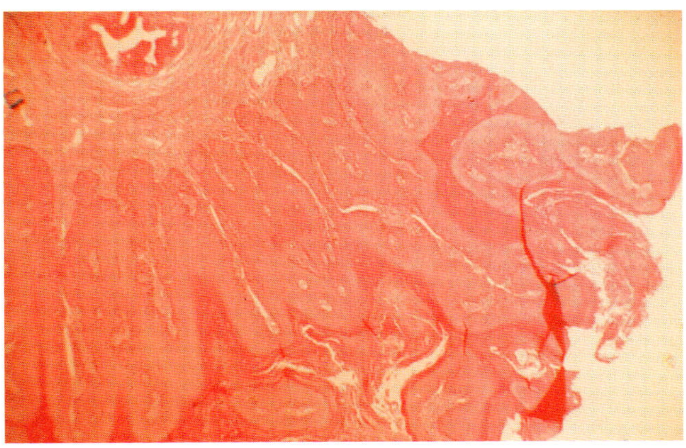

Figure 6.16 Histological section of the lesion shown in Figures 6.14 and 6.15 shows a verrucous carcinoma. In this case the cytology led to underdiagnosis, as it is not possible to give a specific diagnosis of malignancy on cytobrushings of this lesion. The full clinical description of the lesion was not stated on the request form. ×40.

BENIGN PAPILLOMA OF THE ORAL MUCOSA

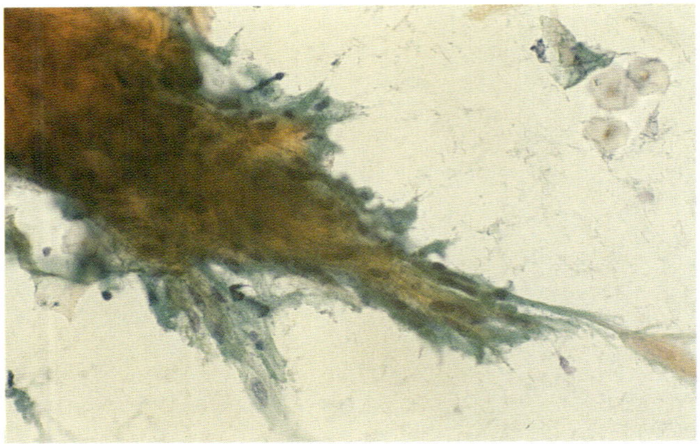

Figure 6.17 An exfoliated sheet of cohesive, spindle-shaped squamous cells with enlarged nuclei, but with plentiful cytoplasm, in a brushing from a papillary lesion. There is mild squamous atypia, but no evidence of dysplasia. ×100.

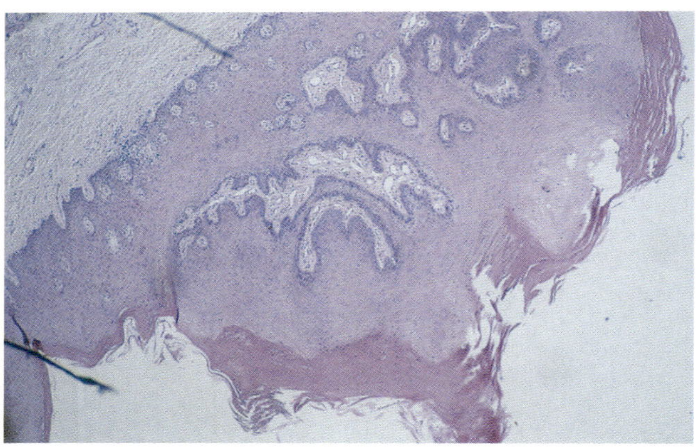

Figure 6.18 Histological biopsy of the lesion shown in Figure 6.17, showing a benign squamous papilloma. ×40.

SEVERE DYSPLASIA/CARCINOMA IN SITU OF THE HARD PALATE

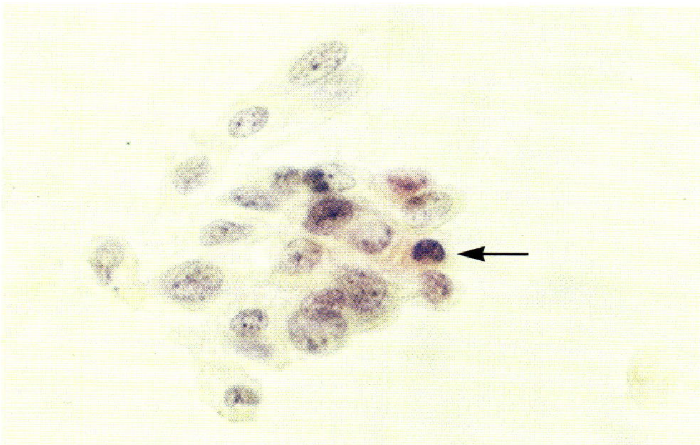

Figure 6.19 Brushings from a lesion on the hard palate. A group of cells showing features of severe dysplasia: apoptosis is noted (arrow). Chromatin is disturbed and nucleoli are enlarged; however, there is overlapping of nuclei. Severe dysplasia with suspicion of invasion was diagnosed. ×400.

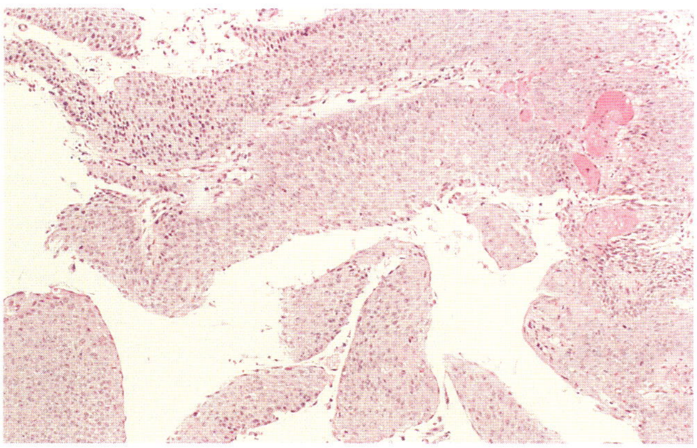

Figure 6.20 The first histological biopsy was fragmented and showed severe dysplasia/carcinoma in situ. Invasion could not be assessed. ×40.

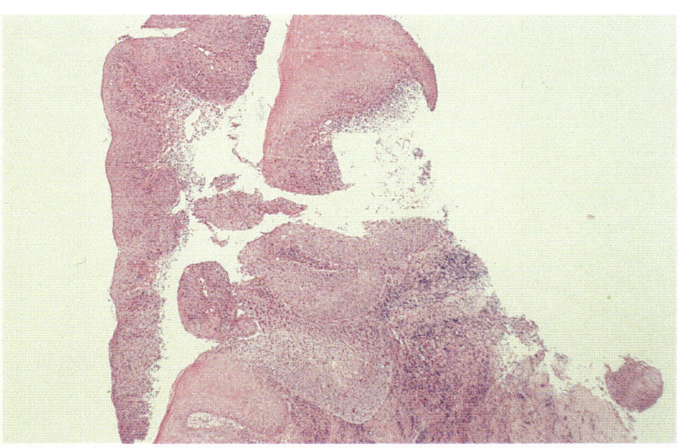

Figure 6.21 The second fragmented biopsy was diagnosed as severe dysplasia/carcinoma in situ with no invasion. However, close clinical and cytological follow-up was recommended. ×40.

SPINDLE CELL SQUAMOUS CARCINOMA

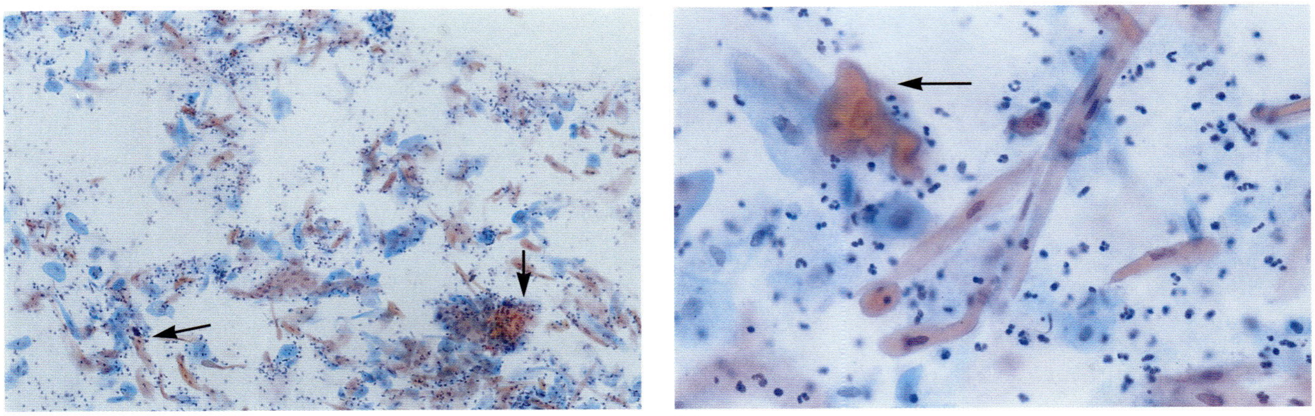

(a) (b)

Figure 6.22 (a) and (b) Cytobrushings obtained from a shallow ulcerating lesion in the alveolar sulcus. Highly atypical loose-lying squamous cells, keratinizing and non-keratinizing, are shown, with long cytoplasmic tails and abnormal nuclei. Some of the nuclei in the spindle-shaped cells are elongated. An apparent keratin pearl is shown on the high-power view. These cells raise suspicion of a squamous carcinoma, possibly spindle cell in type. ×100 and ×400.

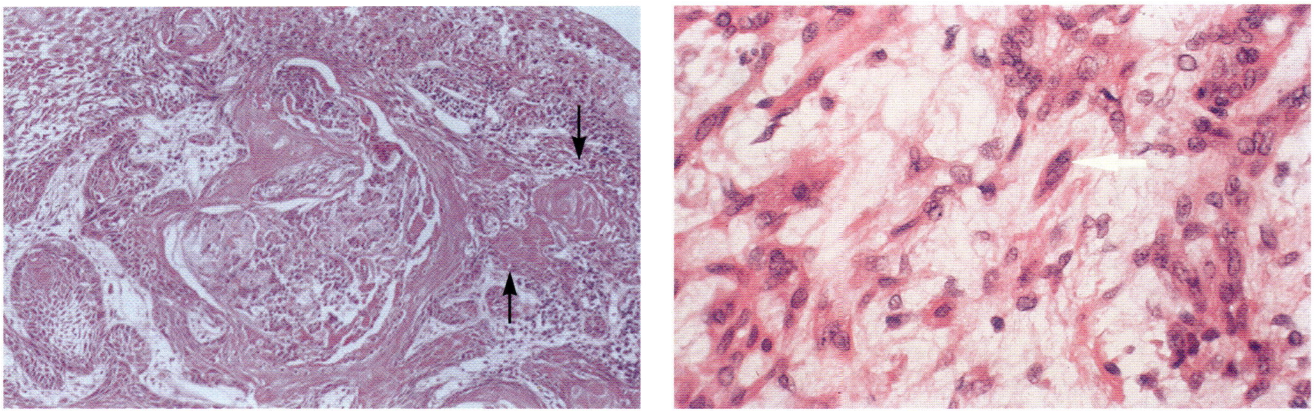

(a) (b)

Figure 6.23 (a) and (b) Biopsy of the lesion shown in Figure 6.22 confirmed an invasive keratinizing squamous carcinoma, with a predominant spindle cell component. Note the keratin pearls and the spindle-shaped cells. ×100 and ×400.

CONTAMINANTS IN ORAL SPECIMENS

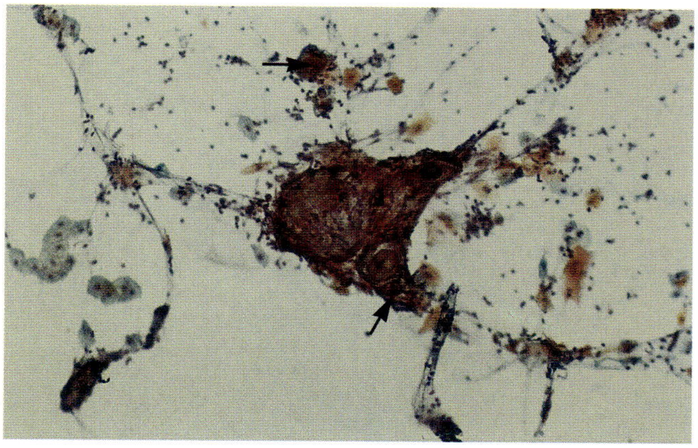

Figure 6.24 Low-power view of oral brushings from a patient with obvious carcinoma and poor oral hygiene. Note particles of vegetable material, probably cellulose. ×100.

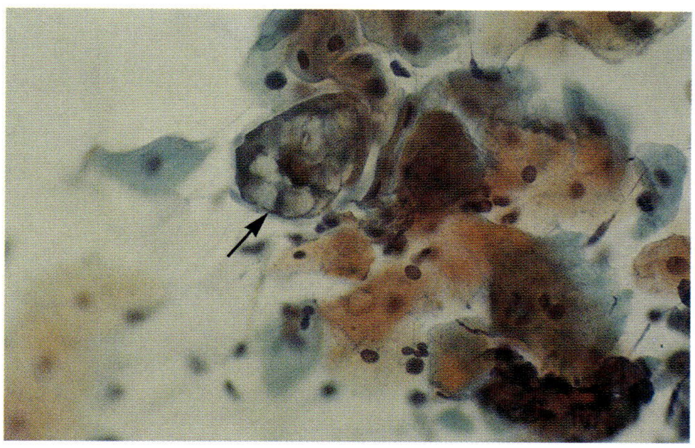

Figure 6.25 A cluster of dysplastic squamous cells is shown together with a food particle (cellulose) in this brushing from a patient with advanced oral carcinoma and poor oral hygiene. ×400.

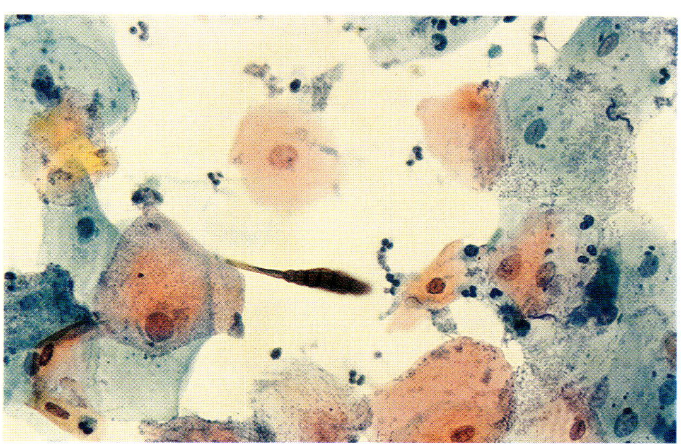

Figure 6.26 Contaminant from tap water in an oral brushing: the non-pathogenic fungus, *Alternaria* spp. Its shape has been fancifully likened to a hand-grenade. Note the normal oral flora associated with the normal squamous cells. ×400.

ADENOCARCINOMA OF THE HARD PALATE

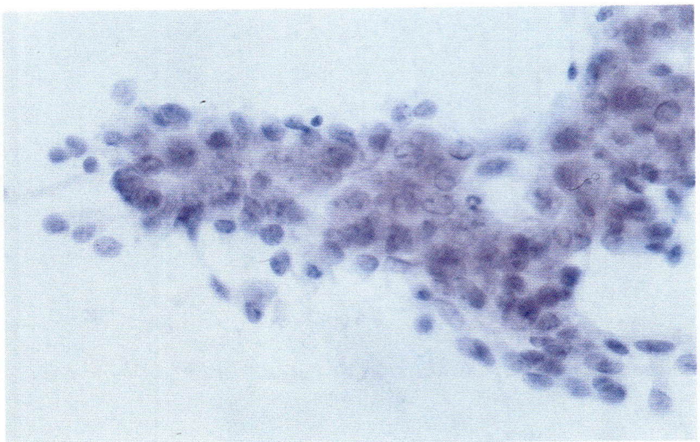

(a)

Figure 6.27 (a) FNA of lesion on the hard palate. Clusters of highly atypical crowded pleomorphic cells are seen. Cytoplasm is scanty and nucleoli are enlarged and pleomorphic. Differentiation is uncertain. ×400.

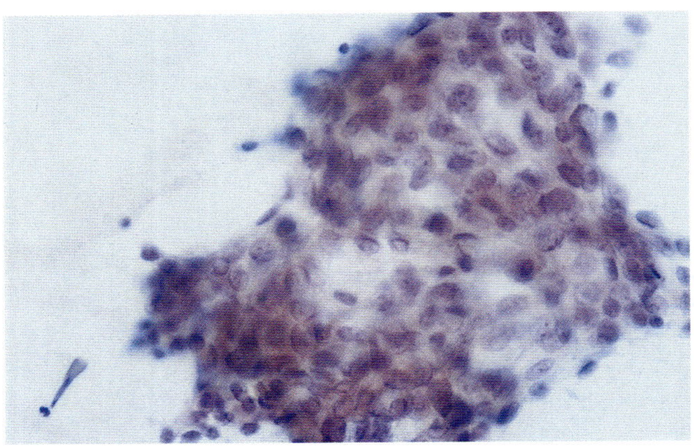

(b)

Figure 6.27 (b) FNA of the same lesion as shown in Figure 27(a). Clusters of crowded pleomorphic nuclei, many with small, but prominent nucleoli. The cells have a three-dimensional arrangement. Differentiation is uncertain. ×400.

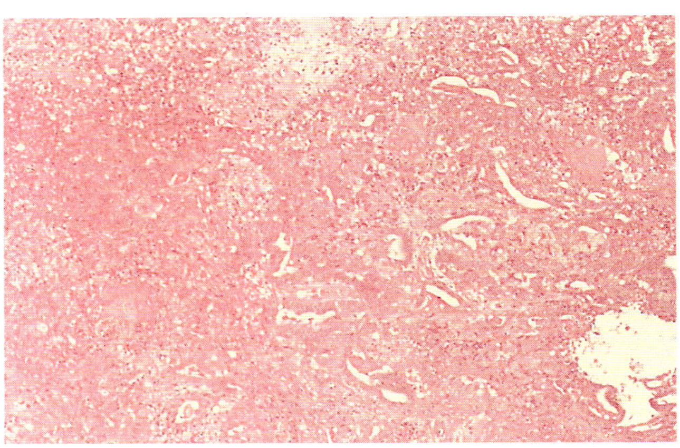

Figure 6.28 Histological section of the lesion in the hard palate was described as a terminal duct adenocarcinoma, possibly arising from a minor salivary gland. Not seen in this photomicrograph is the overlying squamous epithelium, which was ulcerated and dysplastic. ×100.

ORAL ACTINOMYCETES

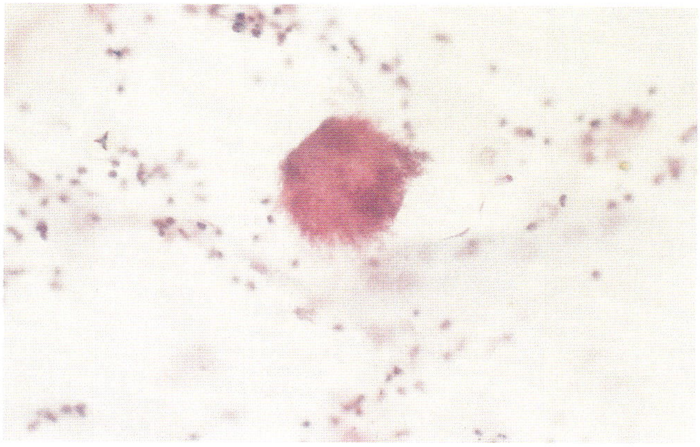

Figure 6.29 Actinomycete-like organisms seen on an oral brushing from a patient with oral carcinoma. This structure is popularly known as a 'purple hedgehog' (see also Figure 7.24, page 100). ×400.

MUCORMYCOSIS OF THE PALATE

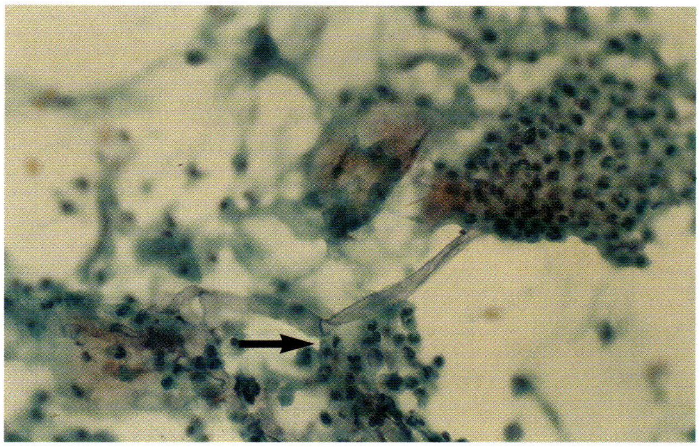

Figure 6.30 Cytobrushing from a destructive lesion of the palate in a comatose patient. Broad, ribbon-like hyphae with septae and non-parallel borders, admixed with acute inflammatory exudate and cellular debris are shown. The patient was diabetic. This condition is also seen in immunosuppressed and debilitated patients. ×100.

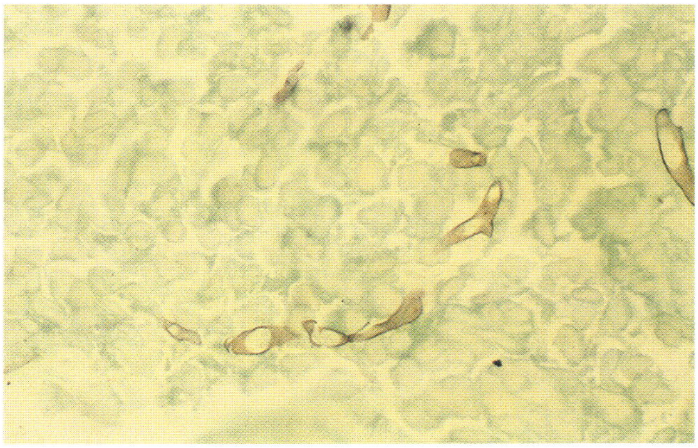

Figure 6.31 Histological section of the lesion stained with methenamine silver to demonstrate the broad hyphae of *Mucor* lying in necrotic debris. ×100.

REFERENCES AND FURTHER READING

Oral brushings

Erenmemisoglu, A., Ustun, H. and Kartal, M. (1995) Carcinoma of the buccal mucosa in smokeless tobacco users: a preliminary study of the use of cytology for early detection. *Cytopathology* **6**: 403–408.

Jones, A. C., Pink, F. E., Sandow, P. L. *et al.* (1994) The cytobrush plus cell collector in oral cytology. *Oral Surg. Oral Med. Oral Pathol.* **77**: 95–99.

Lau, S. K., Hsu, C. S., Sham, J. S. T. and Wei, W. I. (1991) The cytological diagnosis of nasopharyngeal carcinoma using a silk swab stick. *Cytopathology* **2**: 239–246.

Learmonth, G. M., Sellars, S., Duff, J. and Stannard, C. (1990) Cytopathology of the head and neck as an aid to the management of ENT patients. (Abstract.) *Proc. Swed. Soc.* **2**: 143–144.

Ogden, G. R., Cowpe, J. G. and Green, M. (1992) Cytobrush and wooden spatula for oral exfoliative cytology. *Acta Cytol.* **36**: 706–710.

Reddy, C. R. R. M., Kameswari, V. R., Prahlad, D. and Ramulu, C. (1977) Correlative study of exfoliative cytology and histopathology of oral carcinomas. *J. Oral Surg.* **33**: 435–438.

Silverman, S., Bilimoria, K. F., Bhargava, K., Mani, N. J. and Shah, R. A. (1977) Cytologic, histologic and clinical correlations of precancerous and cancerous oral lesions in 57,519 industrial workers in Gujarat, India. *Acta Cytol.* **31**: 196–198.

Mucormycosis

Litzsky, L. and Schwartz, A. M. (1987) Intraoperative cytodiagnosis of paranasal sinus mucormycosis in an immunocompetent host. *Acta Cytol.* 31: 533–534.

Travis, W. D. and Wold, L. E. (1987) Immunoperoxidase staining of fine-needle aspiration specimens previously stained by the Papanicolaou technique. *Acta Cytol.* **31**: 517–520.

7 Nasopharynx, tonsils and larynx

INTRODUCTION

The nasopharynx is the cavity lying behind and above the nasal choanes and the soft palate. The lateral angles of the nasopharyngeal cavity extend to the fossa of Rosenmuller, and incorporate lymphoid tissue. The nasopharynx is covered by ciliated respiratory epithelium, which undergoes squamous metaplasia and dysplasia in situations of chronic irritation and inhalation of toxic substances. Many neoplasms of the nasopharynx are undifferentiated, giving rise to the term 'lymphoepithelioma'.

The tonsils consist of collections of lymphoid tissue covered by squamous epithelium. They may be the site of carcinomas, lymphoproliferative processes and lymphomas. The larynx and surrounding structures are covered by squamous epithelium.

Diagnosis of lesions of the nasopharynx, tonsils and sinuses is often difficult owing to the difficulty of visualization while procuring a biopsy. The straight-handled cytobrush is excellent for collecting specimens from easily accessible lesions; however, the long-handled flexible endobrush has been found to provide excellent, very cellular specimens from inaccessible lesions in the larynx, nasopharynx and epiglottis. A retractable transparent sheath covers the distal end of the instrument to avoid cell pick-up before the brush reaches the lesion. This sheath can be retracted when the operator wishes to sample a lesion or suspicious area. The brush can then be recovered by the sheath during removal, to prevent loss of sampled cellular material. Often a good representative cytobrushing will provide more information than small, friable, fragmented pieces of tissue.

NASOPHARYNGEAL BRUSHINGS

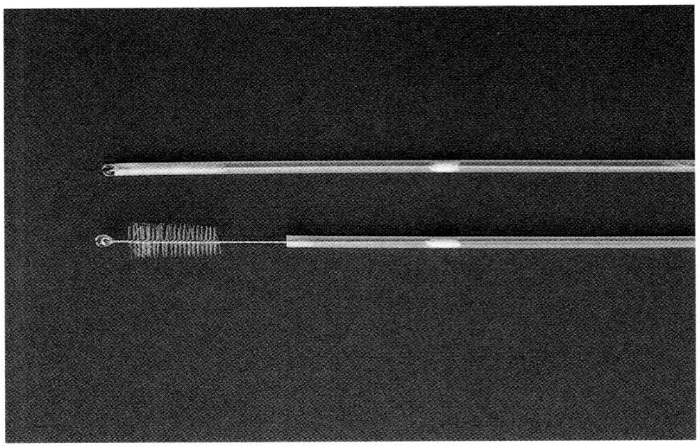

Figure 7.1 Selective brushings can be taken from the midline and the lateral walls using the flexible endobrush. In this way the clinician can pinpoint the focus of a small neoplastic lesion more precisely, and radiotherapy can be more accurately directed, lessening the area exposed to irradiation.

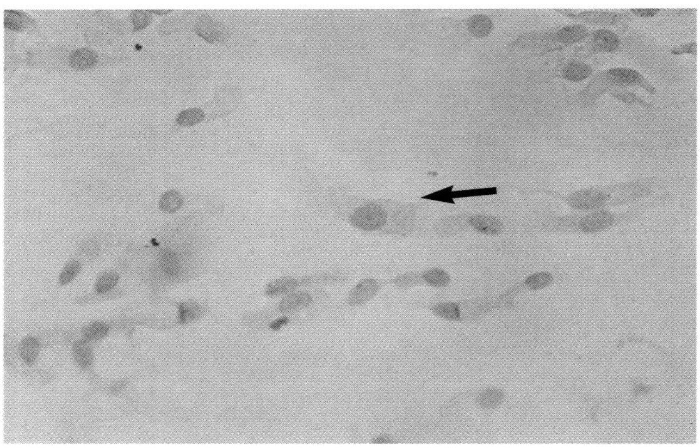

Figure 7.2 Normal ciliated columnar epithelial cells from the postnasal space. A single normal, eosinophilic angular squamous cell is also present. Note the basal orientation of the nuclei, some of which are uniformly enlarged and have small, prominent pinpoint nuclei. These are benign reactive changes. ×100

SQUAMOUS CARCINOMA IN THE NASOPHARYNX

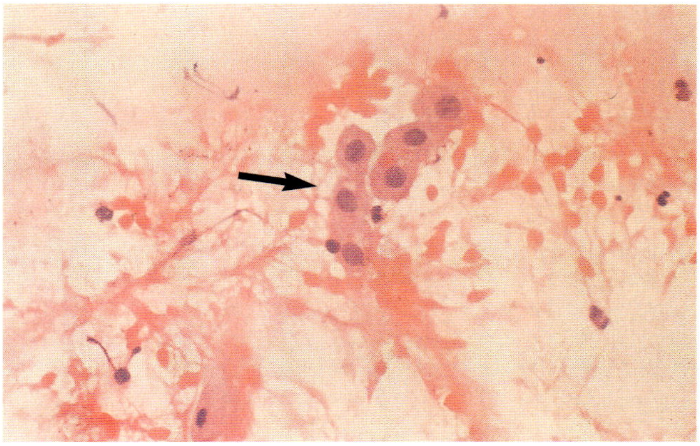

Figure 7.3 Brushing from a patient with an ulcerating lesion in the nasopharynx: metaplastic squamoid cells with uniformly enlarged nuclei are shown. The nuclear outlines are regular. No nucleoli are present. There is blood in the background. ×100.

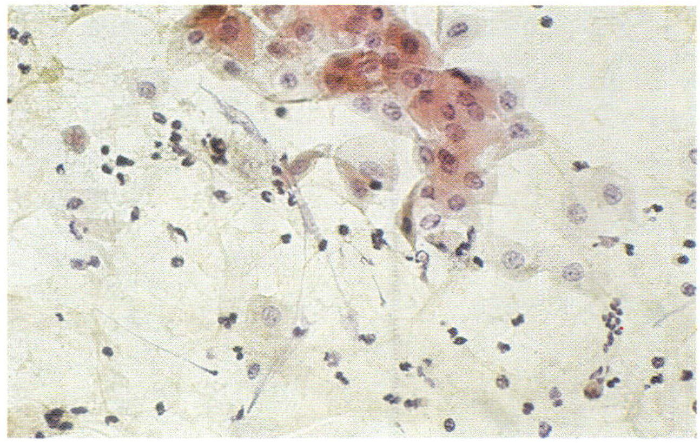

Figure 7.4 This photomicrograph shows a sheet of atypical squamoid cells with a few spindle-shaped squamoid cells lying loose in the background. ×100.

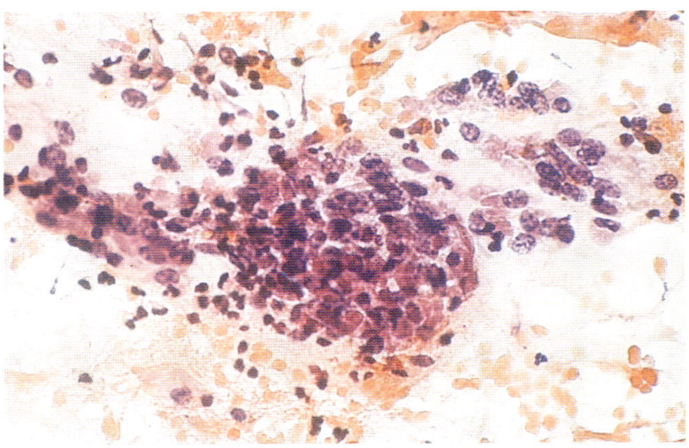

Figure 7.5 Highly atypical cells with enlarged nuclei and scanty cytoplasm, lying in a background of blood. Many of the atypical cells are obscured by lymphoid cells and polymorphs. The nuclei are crowded and overlapped. Nuclear chromatin is disturbed. These cells raise suspicion of a carcinoma. ×100.

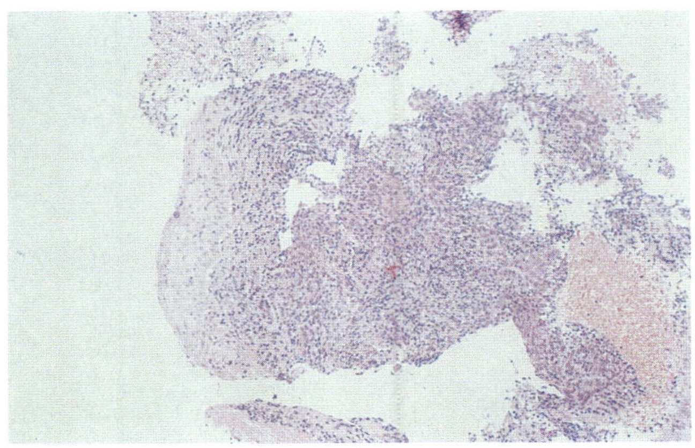

Figure 7.6 Histological section of a fragmented biopsy of the lesion shown in Figures 7.3–7.5 confirms the presence of an invasive squamous carcinoma, which is infiltrated by inflammatory cells. ×100.

NASOPHARYNGEAL CARCINOMA

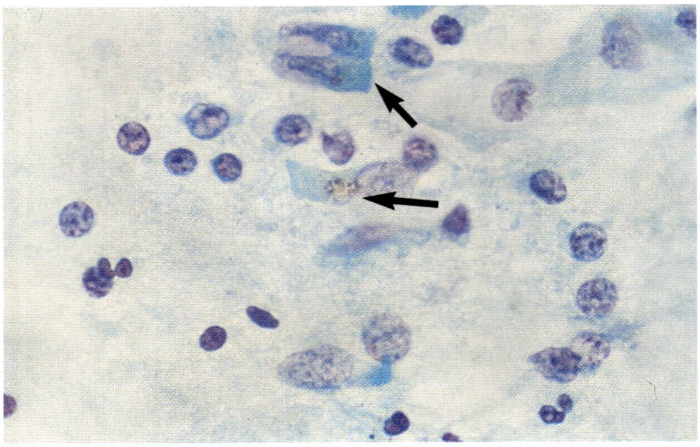

Figure 7.7 Endobrush sample of nasopharyngeal lesion. The photomicrograph shows poorly differentiated, round cells with altered N/C ratio, scanty or no cytoplasm, large round nuclei with disturbed chromatin and small eosinophilic nucleoli. These are compatible with an undifferentiated nasopharyngeal carcinoma (NPC). Markedly reactive columnar cells are also present (arrows). The cytoplasm of these is plentiful, the cell shape is within normal limits, the nuclei are enlarged, but they are regular in size and shape. Nuclear chromatin is pale. Nucleoli are not prominent. These reactive changes are often seen adjacent to neoplastic lesions. ×400.

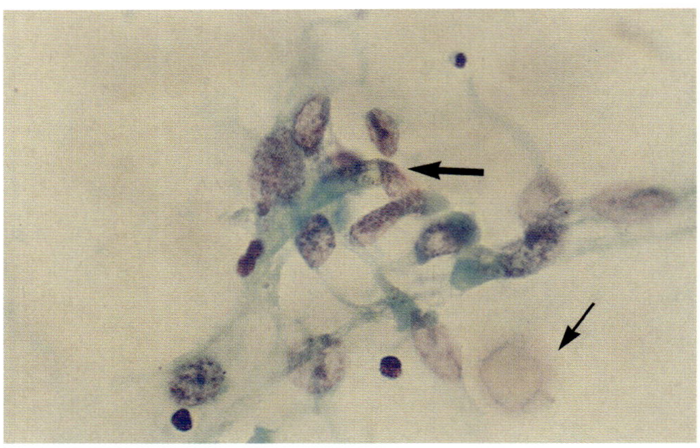

Figure 7.8 Brushings from a nasopharyngeal lesion in a patient with severe unexplained weight loss. Large pleomorphic pale cells with disturbed nuclear chromatin and variable pinpoint nucleoli. There is loss of intercellular cohesion. Degenerative changes are present. These cells raise suspicion of an undifferentiated NPC. A talc granule is present at the small arrow. ×400.

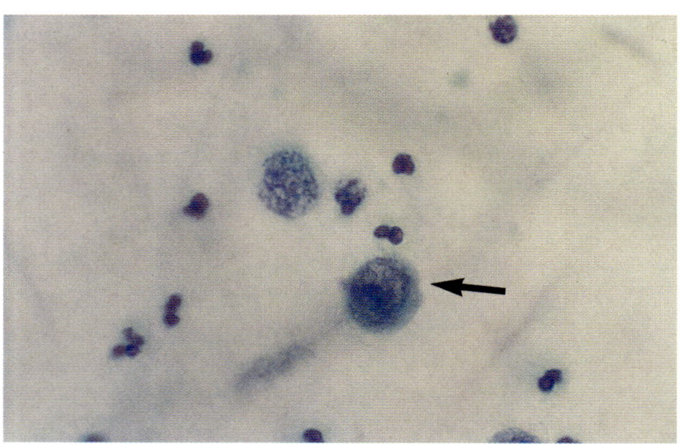

Figure 7.9 High-power view of single-lying pleomorphic undifferentiated cells from NPC. Note the pale nuclei, disturbed chromatin and pinpoint nucleoli. ×400.

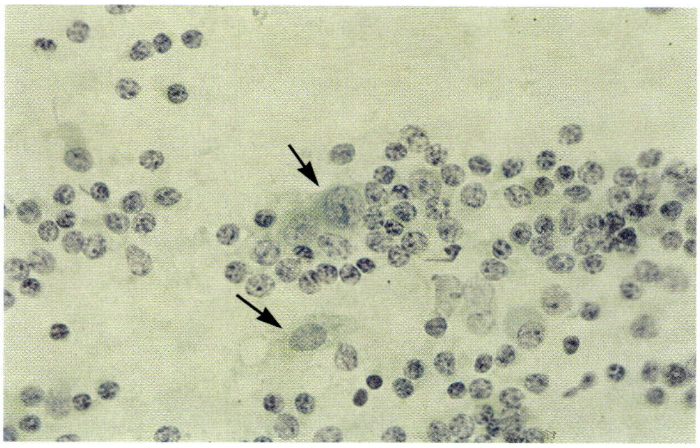

Figure 7.10 Brushing from postnasal space. Large, pale, poorly differentiated cells (arrows) lying in a background of smaller lymphoid cells. These raise a suspicion of NPC with a predominant lymphoid component, or a high-grade lymphoma. A confirmatory biopsy is mandatory in such cases. ×100.

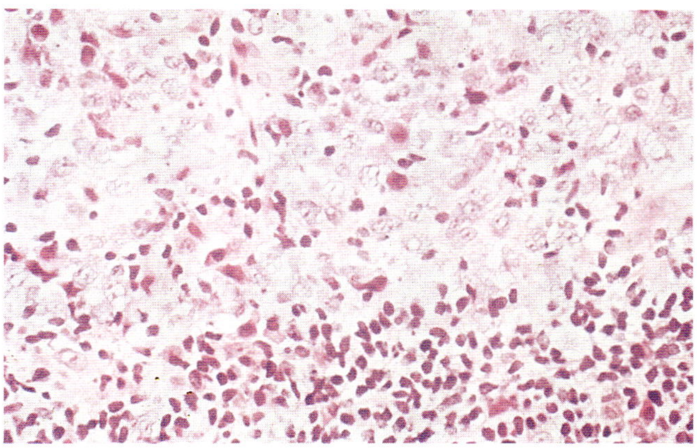

Figure 7.11 Histological section of the suspected NPC shows a neoplasm composed of small, pale, undifferentiated cells with prominent nucleoli. Note the mantle of adjacent lymphoid cells at the bottom of the illustration. ×100.

OCCULT NASOPHARYNGEAL SQUAMOUS CARCINOMA

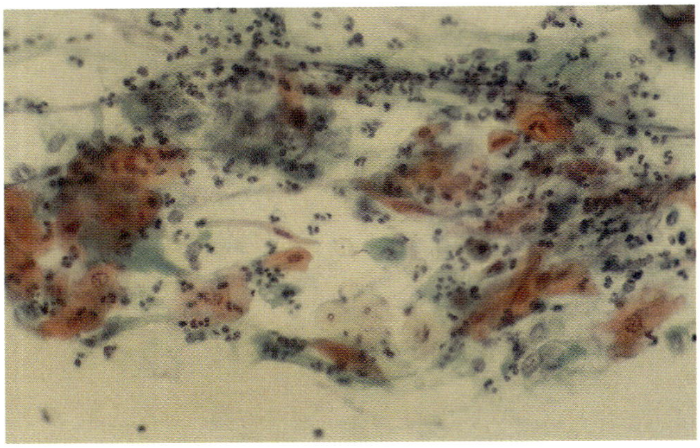

Figure 7.12 Nasopharyngeal brushing from an elderly male who presented with huge metastatic neck nodes. All efforts to locate a primary lesion failed, including careful endoscopic examination of the nasopharynx, larynx and epiglottis. Brushings were done selectively from the lateral walls and the midline. This photomicrograph shows highly atypical keratinized and non-keratinized squamous cells, suggestive of squamous carcinoma, which were found on the brushing from the left lateral wall. ×100.

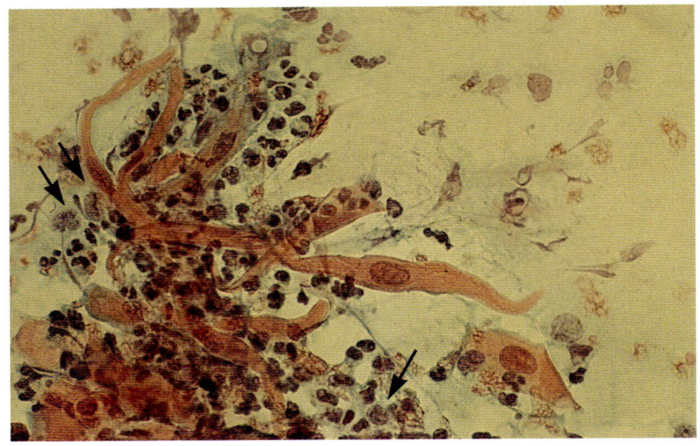

Figure 7.13 High-power view of bizarre, malignant keratinized squamous cells seen in a nasopharyngeal brushing. Note the atypical lymphoid cells of varying degrees of maturity (arrows) intermixed with the maligant squamous cells. ×400.

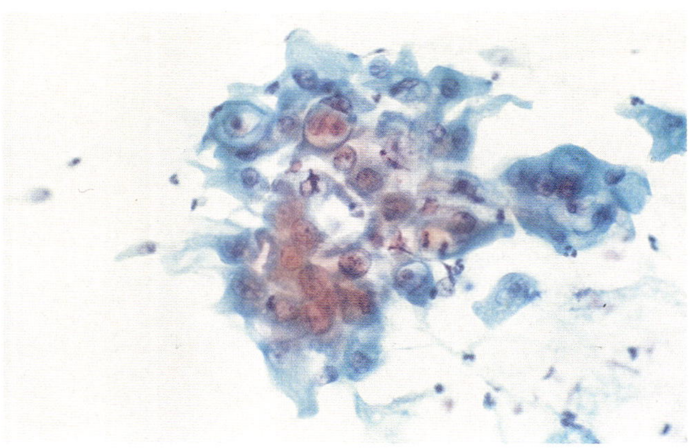

Figure 7.14 Non-keratinized malignant squamous cells from the same specimen as shown in Figure 7.13. ×400.

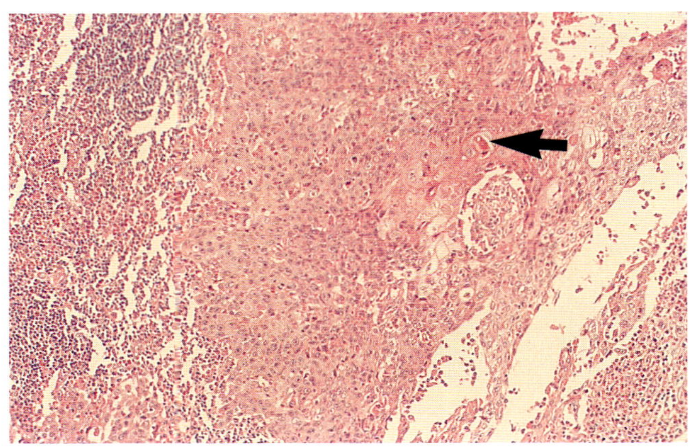

Figure 7.15 Histological section of a neck node showing metastatic keratinizing squamous carcinoma, probably from the nasopharyngeal lesion shown in Figures 7.12–7.14. Note the squamous pearl (arrow). ×100.

EPIGLOTTIS – KERATINIZING SQUAMOUS CARCINOMA

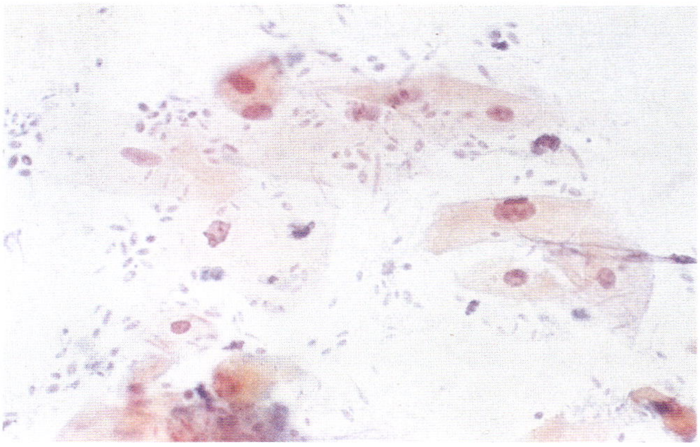

Figure 7.16 Endobrushing of lesion on the epiglottis showing squamous atypia with *Candida* spores and short hyphae. This degree of atypia is not usually seen with *Candida* infection alone. Furthermore, fungal infections within lesions of the oropharynx should arouse suspicion of adjacent carcinoma, and the clinician should be alerted to this possibility (see also Figures 6.1b and 6.2, page 78). ×100.

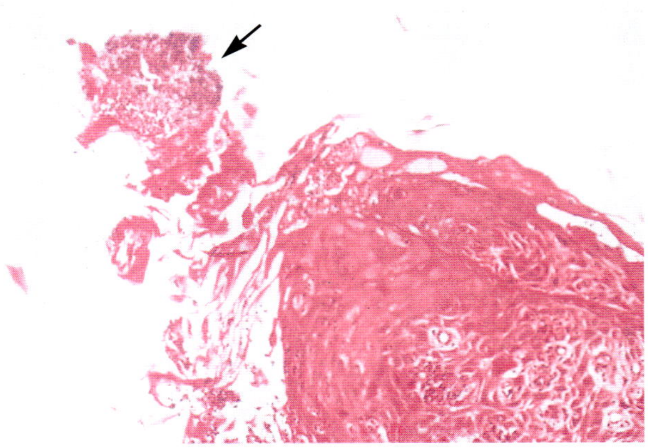

Figure 7.17 Histological biopsy of the lesion illustrated in Figure 7.16 confirmed the presence of a keratinizing squamous carcinoma and associated 'ball' of *Candida* spores (arrow). ×100.

TONSILS

Squamous carcinoma and lymphoma may present as masses in the tonsils. Cytobrushings may help to differentiate these lesions for triage purposes. In patients who present clinically with enlarged lymph nodes in the neck, endobrush cytology may help to suggest a primary lesion in the tonsils, thus speeding up the final diagnostic process and referral for treatment. An attempt to classify the grade of lymphoma is not really necessary as this will be done subsequently on the histology sections (see Chapter 5). However, the pathologist can be alerted to the possibility of a lymphoma, prior to surgical biopsy. Arrangements for tissue imprints and molecular biological studies can be made, thus ensuring optimum diagnostic material.

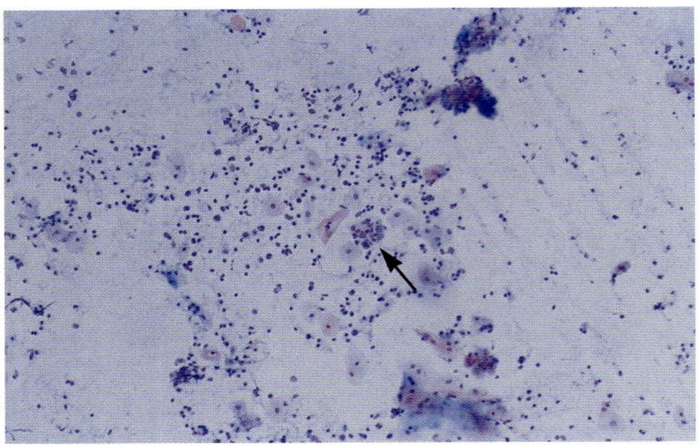

Figure 7.18 Cytobrushing of tonsillar lesion showing predominantly lymphoid cells and scattered, normal loose-lying squamous cells. Note the follicular arrangement of some lymphoid cells (arrow). ×100.

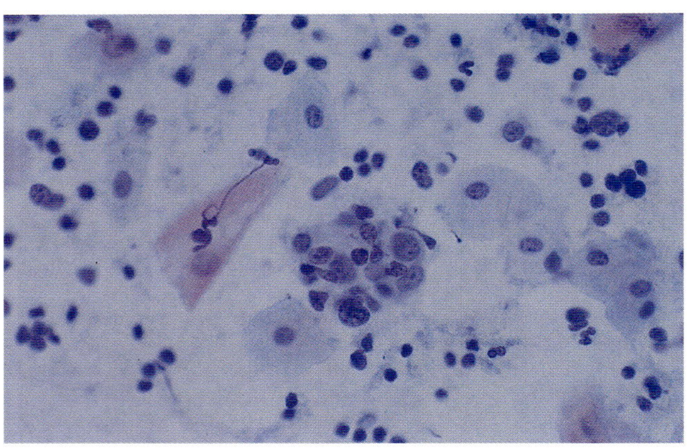

Figure 7.19 High-power view of cells seen in Figure 7.18. The lymphoid cells in the follicular cluster have disturbed chromatin, pleomorphic nuclei and irregular nuclear outlines. No tingible-body macrophages are seen. Note the normal squamous cells. These appearances suggest a lymphoid neoplasm rather than a carcinoma. ×400.

LYMPHOMA IN TONSILS

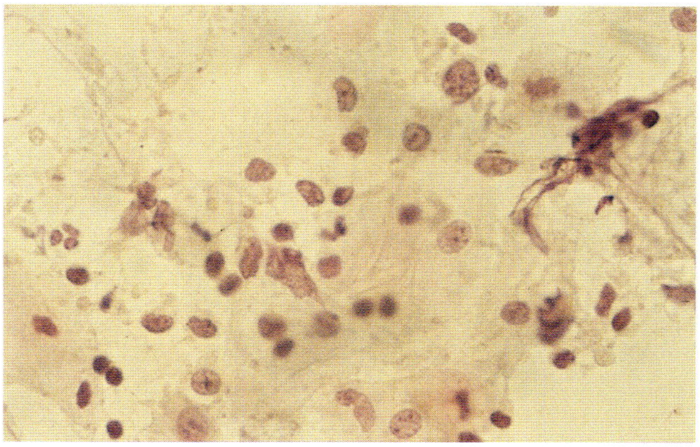

Figure 7.20 Tonsillar brushing showing single-lying, highly atypical lymphoid cells with disturbed chromatin and prominent small nuclei, suggestive of lymphoma. ×400.

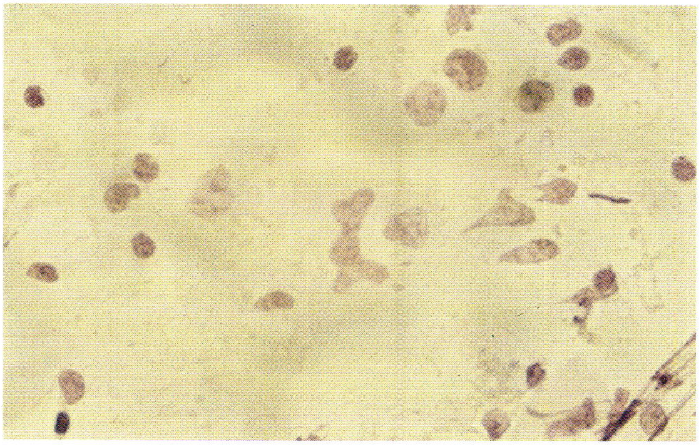

Figure 7.21 A particularly bizarre lymphoid cell in a brushing from the same specimen as seen in Figure 7.20. This shape is thought to be due to the 'unfolding' of a cleaved nucleus. ×400.

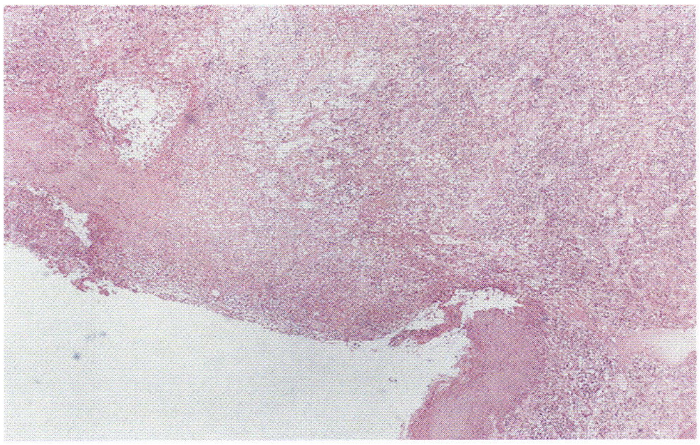

Figure 7.22 Histological section of the ulcerating tonsillar lesion described in Figures 7.20 and 7.21 confirmed the presence of a high-grade lymphoma in the tonsil. ×100.

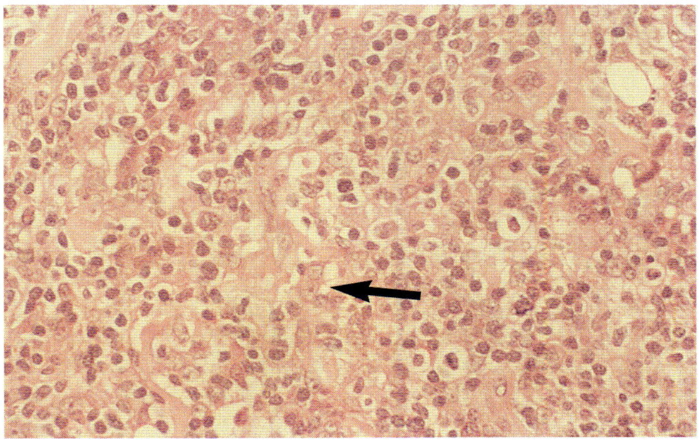

Figure 7.23 High-power view of Figure 7.22. ×400.

ACTINOMYCOSIS IN THE TONSIL

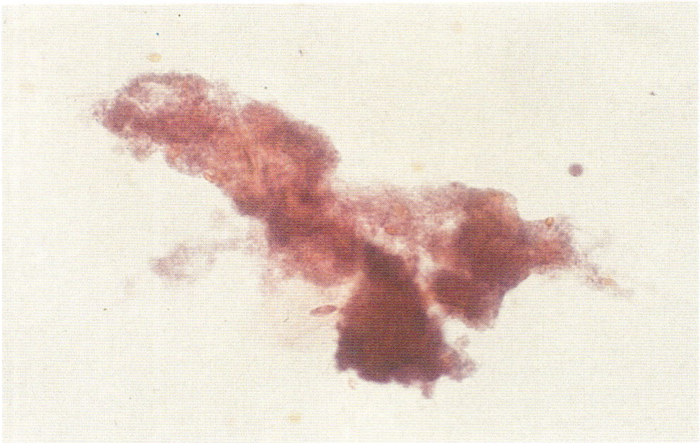

Figure 7.24 Brushing from the base of tongue close to the tonsil. Colonies of *Actinomyces* are often seen in brushings adjacent to a neoplastic lesion (see also Figure 6.24, page 89). ×100.

WELL-DIFFERENTIATED CARCINOMA OF THE TONSIL

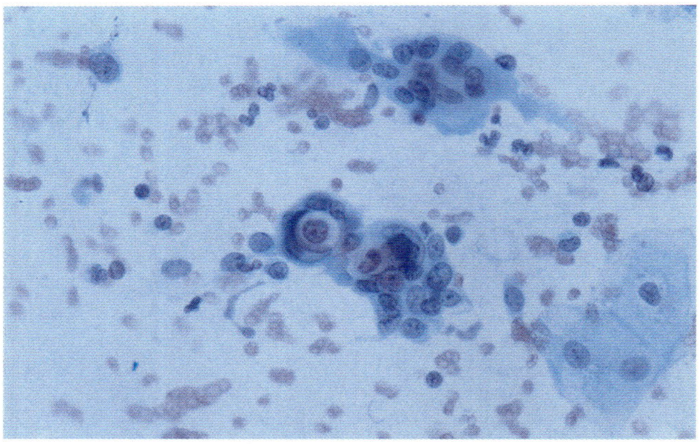

(a)

Figure 7.25 (a) This brushing from an ulcerating lesion on the tonsil shows highly atypical loose-lying squamoid cells in a background of erythrocytes. Note cell in cell arrangement, often seen in smears from superficial squamous carcinoma.

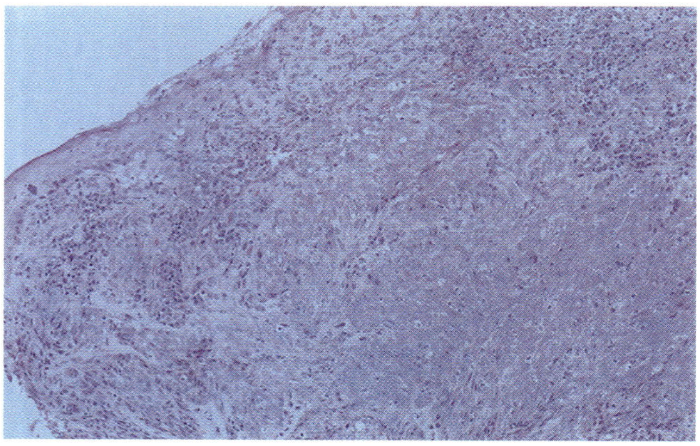

(b)

Figure 7.25 (b) Histological section of the lesion shown in Fig 7.25a confirms the presence of a well differentiated squamous carcinoma.

POORLY DIFFERENTIATED CARCINOMA IN THE TONSILLAR AREA

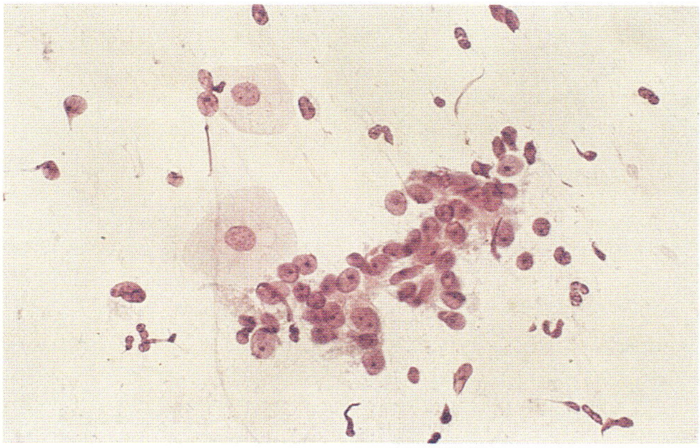

Figure 7.26 A cluster of small undifferentiated cells with pinpoint nucleoli, devoid of cytoplasm. These cells are suspicious of an NPC. Note the benign reactive squamous cells in the background. ×100.

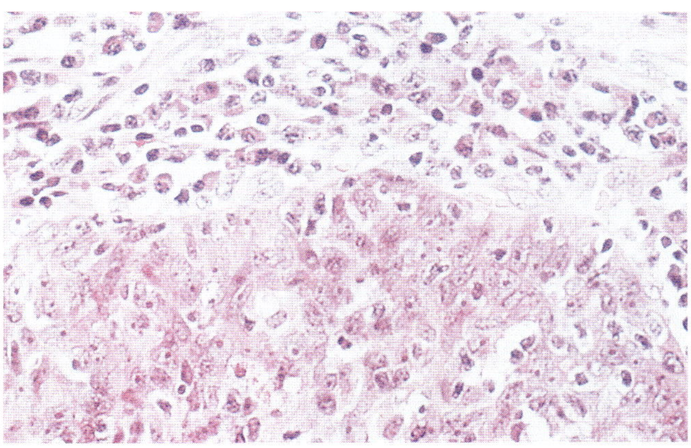

Figure 7.27 Histology of base of tongue/tonsil lesion described in Figure 7.26, showing an undifferentiated neoplasm consisting of clustered small cells with pinpoint nucleoli bordering on lymphoid tissue (at top). Note the adjacent mantle of lymphoid cells and plasma cells at the top of the photomicrograph. ×100.

LARYNX–SQUAMOUS CARCINOMA

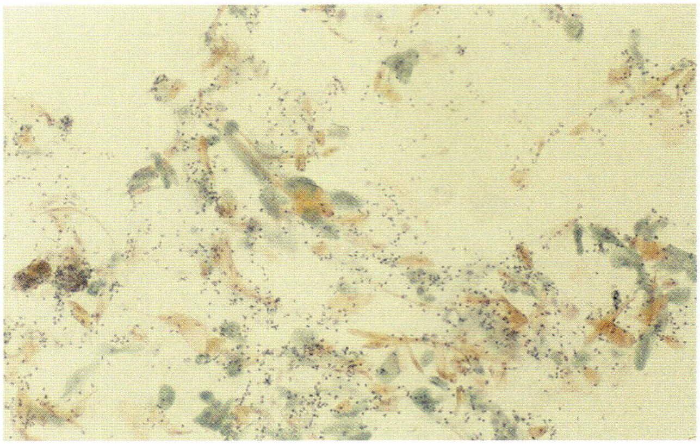

Figure 7.28 Endobrushing of a fungating lesion on the tonsil. Loose lying highly atypical keratinized squamous cells cause suspicion of a squamous carcinoma. ×100.

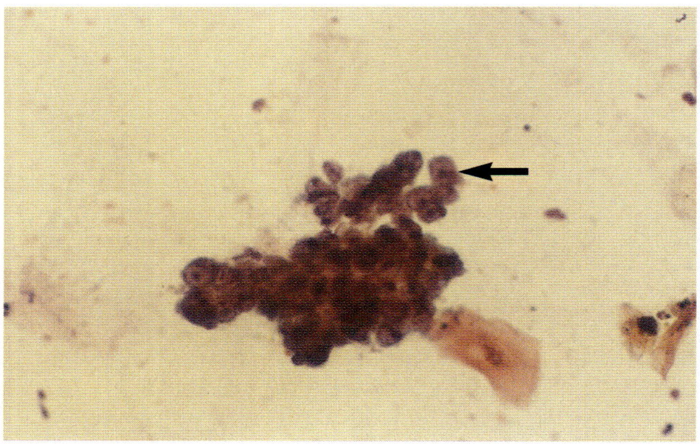

Figure 7.29 Laryngeal brushing: a sheet of highly atypical pleomorphic squamous cells, compatible with severe dysplasia at least. ×400.

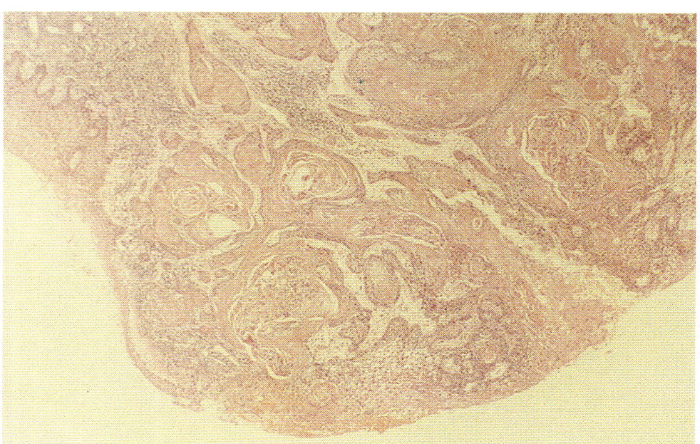

Figure 7.30 Histological biopsy of the lesion on the larynx illustrated in Figure 7.28 showed a well-differentiated squamous carcinoma. ×100.

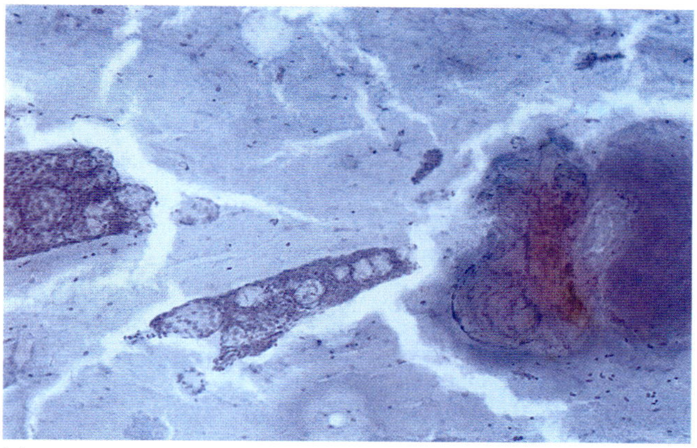

(a)

Figure 7.31 (a) This FNA of a soft parapharyngeal mass yielded cellular tissue with a stromal component and a predominant epithelial component. The latter consisted of epithelial cells grouped around apparent hyaline masses, giving it the distinctive pattern associated with adenoid cystic neoplasms. ×100.

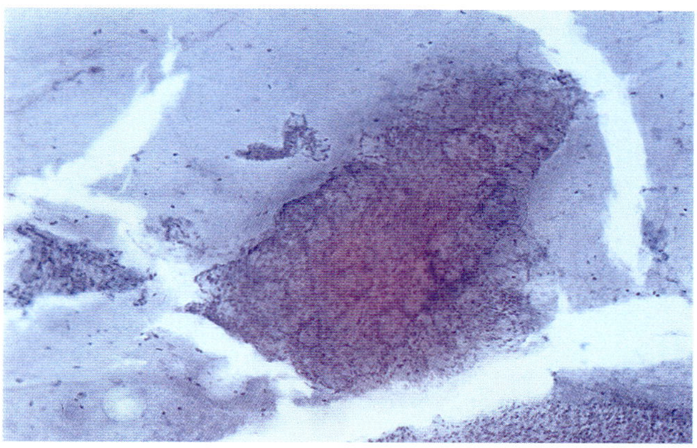

(b)

Figure 7.31 (b) This view of the specimen discussed in Figure 7.31a, shows a slightly different pattern with a rather cribriform arrangement. The differential diagnosis lies between a pleomorphic adenoma and a low grade adenoid cystic carcinoma. The clinical history was that of a slow growing innocuous lesion with no associated pain. The latter feature is a common feature of adenoid cystic carcinoma. (see Chapter 1) ×100.

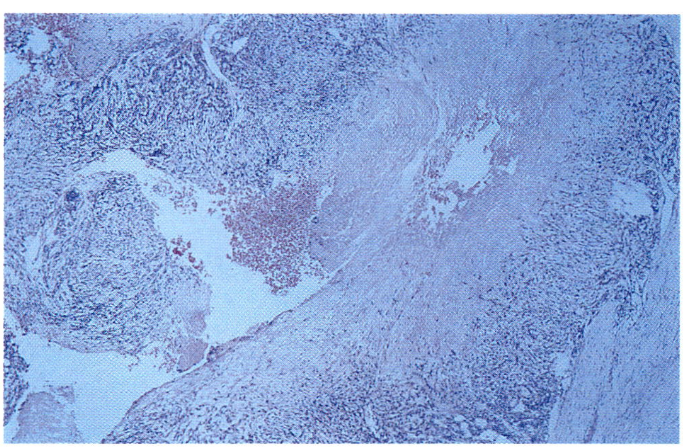

(c)

Figure 7.31 (c) Histological section of the excised mass showed a pleomorphic adenoma with an apparent 'needle track'. In this case the clinical history was more in favour of a benign neoplasm, rather than an adenoid cystic carcinoma. ×100.

FURTHER READING

Chan, M. K. (1990) The value of cytologic examination for nasopharyngeal carcinoma. *Ear, Nose, Throat J.* **69**: 268–271.

Ear, Nose and Throat Journal. (1990) Nasopharyngeal carcinoma. (Review.) *Ear, Nose, Throat J.* **69**: 216–252.

Lau, I., Hsu, C. S., Sham, J. S. T. and Wei, W. I. (1991) The cytological diagnosis of nasopharyngeal carcinoma using a silk swab stick. *Cytopathology* 2: 239–246.

Van Hasselt, A. (1991) *Nasopharyngeal carcinoma*. Hong Kong, Chinese University Press, pp. 157–161.

Warner-Learmonth, G. M., Duff, J. and Sellers, S. (1993) The cytological diagnosis of nasopharyngeal neoplasms using the flexible Endobrush. (Abstract) *Pathol. Res. Pract.* **189**: 840.

8 Nose and paranasal sinuses

INTRODUCTION

Lesions in the paranasal sinuses often present diagnostic problems because of their inaccessibility and the difficulty of obtaining a good histological biopsy. The flexible endobrush yields good cellular material, which can be useful in obtaining a cytological opinion. This in turn guides the clinician in further investigation of the patient, or helps to confirm the spread or recurrence of a neoplasm. Direct imprints can be used intraoperatively. The endobrush can be rinsed in suitable fluid for bacteriological studies.

MULTIPLE MYELOMA

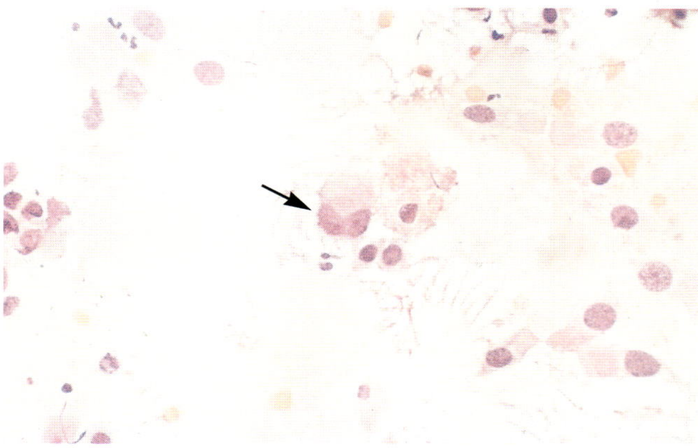

Figure 8.1 Endobrush specimen from a destructive mass in the sphenoid sinus from a patient with a known multiple myeloma in the cervical vertebral bodies. This photomicrograph shows a large plasmacytoid binucleate cell (arrow) in a background of ciliated columnar cells and a few loose-lying, poorly preserved plasmacytoid cells, suggestive of myeloma. ×400.

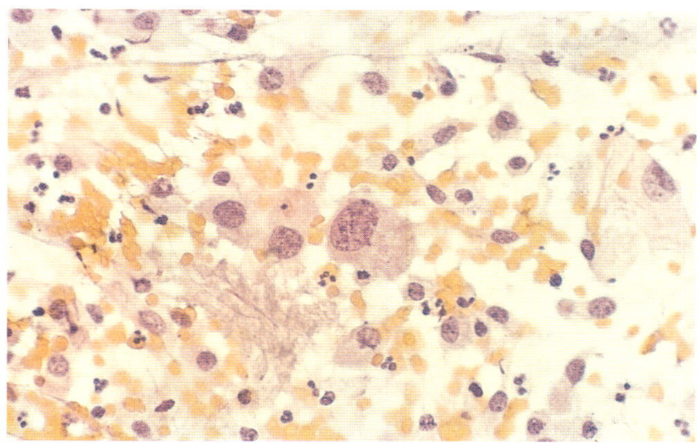

Figure 8.2 Loose-lying large pleomorphic plasmacytoid cells from a myeloma in the sphenoid sinus, shown lying in a background of small plasmoid cells and columnar cells. ×400.

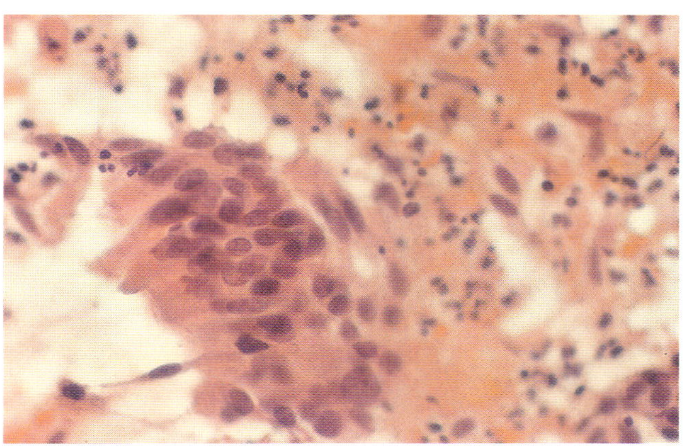

Figure 8.3 A cohesive sheet of markedly reactive columnar epithelial cells from the same brushing. These should not be mistaken for neoplastic cells. Epithelial cells often show marked reactive changes in the presence of a neoplasm. The nuclei are elongated and show pseudopalisading, but the cells are cohesive (see Chapter 7). ×400.

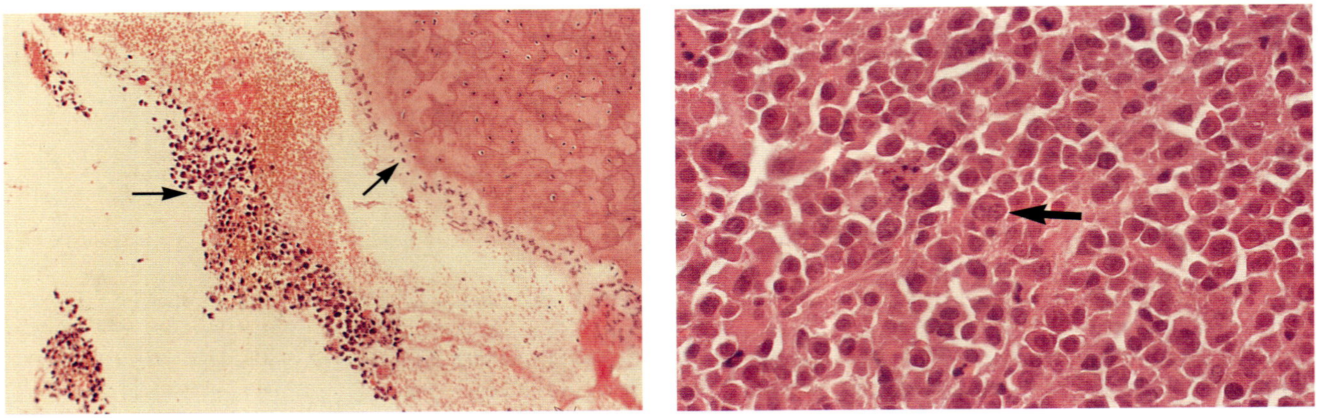

(a) (b)

Figure 8.4 (a) and (b) Sphenoid sinus: Histology of the fragmented biopsy consisted of loose-lying plasmacytoid cells, some of them binucleate (arrow), compatible with a myeloma. Note the proximity of the destructive lesion to bony tissue (arrow). H&E ×100 and ×400.

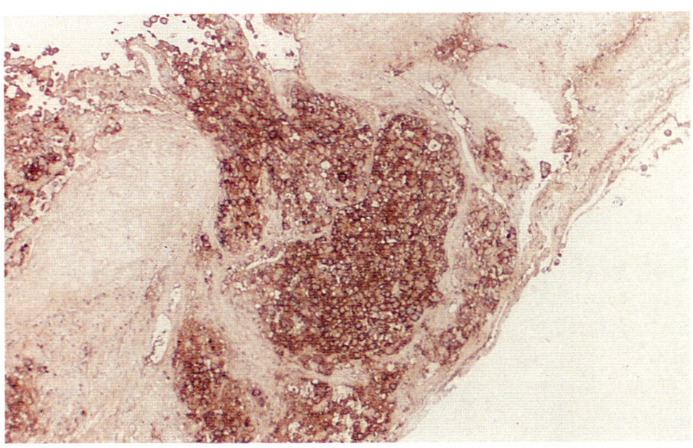

Figure 8.5 Immunocytochemistry of the biopsy confirmed the presence of Lambda-positive cells. These were also seen in the original neoplasm. Lambda. ×100.

MUCORMYCOSIS OF THE NASOSINUSOIDAL AREA

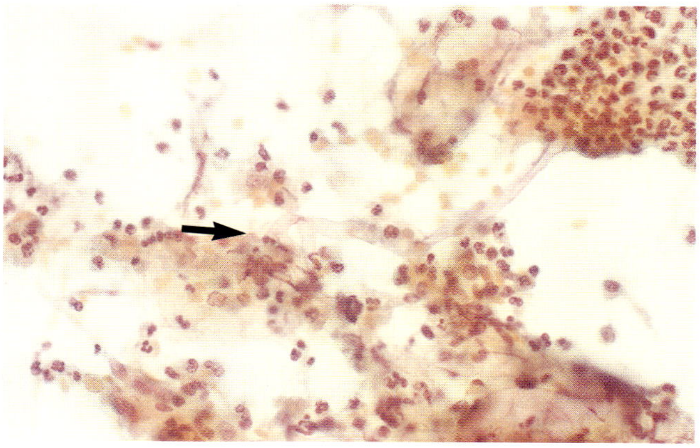

Figure 8.6 Endobrushing from an unidentified comatose patient with a destructive fungating lesion of the palate which extended into the paranasal sinuses. As the patient was considered to be terminally ill, an urgent diagnosis was necessary. Broad hyphae with non-parallel borders were found. The presence of septae is difficult to demonstrate. The fungal hyphae were almost totally obscured by a purulent inflammatory exudate and necrotic debris. These fragile hyphae may be missed in histological sections because only small fragments are present after processing and sectioning, whereas in cytological preparations they remain intact. No atypical squamous or lymphoid cells were found. Mucormycosis was suggested. At the same time diabetes was clinically confirmed. ×400.

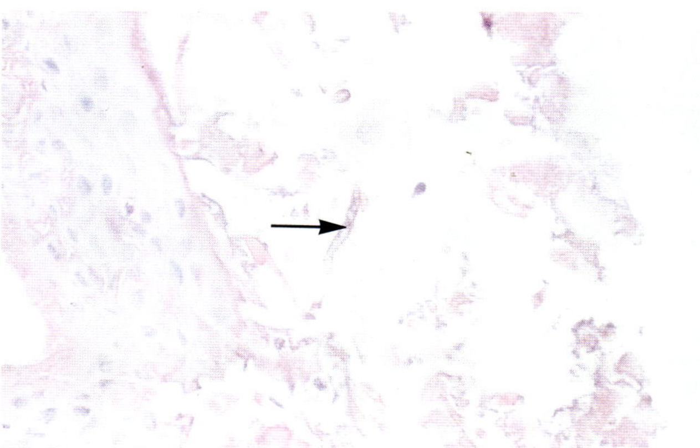

Figure 8.7 Histological section of the degenerate fragmented biopsy showed similar hyphae in a background of destroyed bony tissue and adjacent cartilage, compatible with a diagnosis of mucormycosis. H&E ×100.

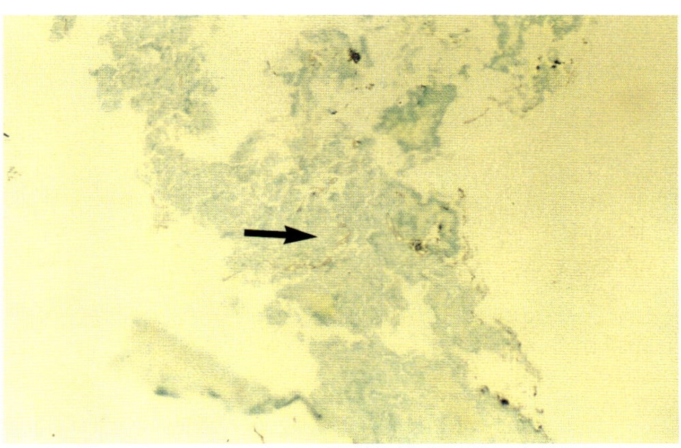

Figure 8.8 Methenamine silver stain highlighted the fragmented fungal hyphae. ×100.

FIBRO-OSSEOUS LESION OF THE PARANASAL SINUSES

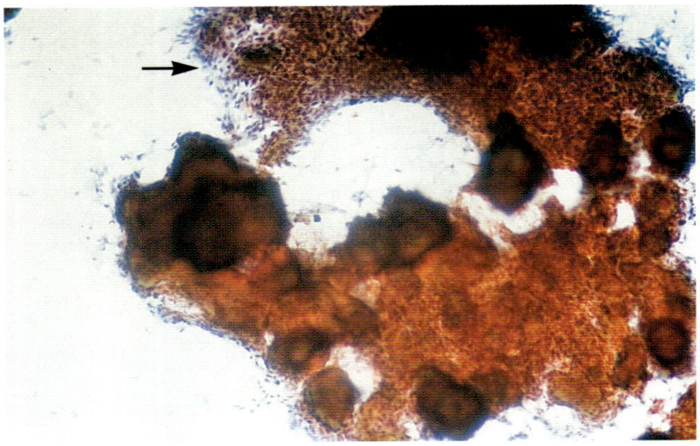

Figure 8.9 Intraoperative smear from a 13-year-old boy who had a mass involving the ethmoid sinus. Progressive unilateral protrusion of the eyeball had been noticed during the previous 6 months. Masses of cohesive tissue with obvious islands of either calcification or bony tissue are present. A very cellular clump of stromal cells is seen (arrow). The clinical differential diagnosis was a 'blast-cell' tumour, a meningioma or a fibro-osseous lesion. The latter are known to display locally aggressive behaviour in prepubertal adolescents (Margo *et al*., 1995). ×40.

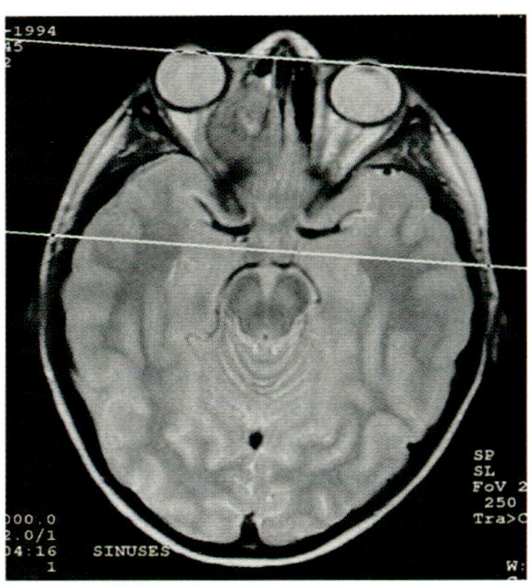

Figure 8.10 MRI scan showing a dense mass obliterating the ethmoid sinus and causing protrusion of the ipsilateral orbit.

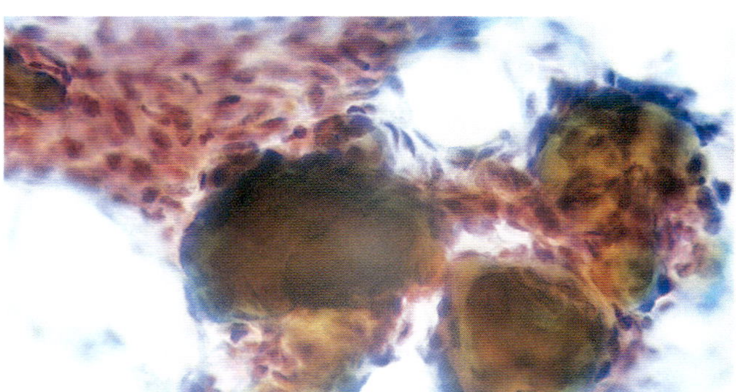

Figure 8.11 Higher magnification of the lesion shown in Figure 8.10: markedly cellular stroma surrounding the round islands of dense material – either dense calcification or bony tissue. The stromal cells are very cohesive. They have scanty pale basophilic cytoplasm and round to oval nuclei. No 'blast' cells are seen. ×100.

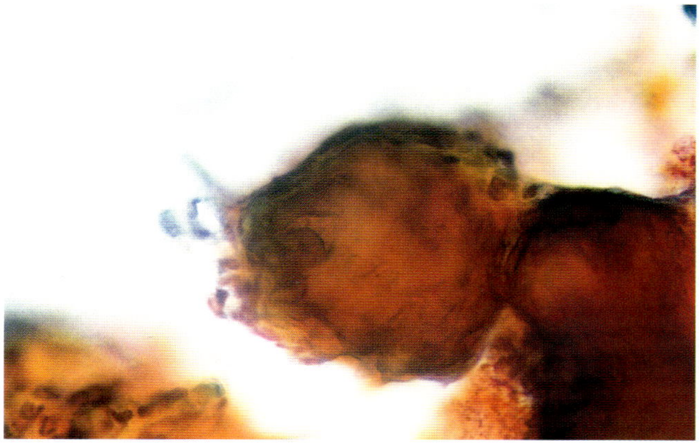

Figure 8.12 A high-power view of the solid 'bony' islands of tissue, showing the lamellar linear contours. The latter are more irregular than those seen in psammoma bodies. ×400.

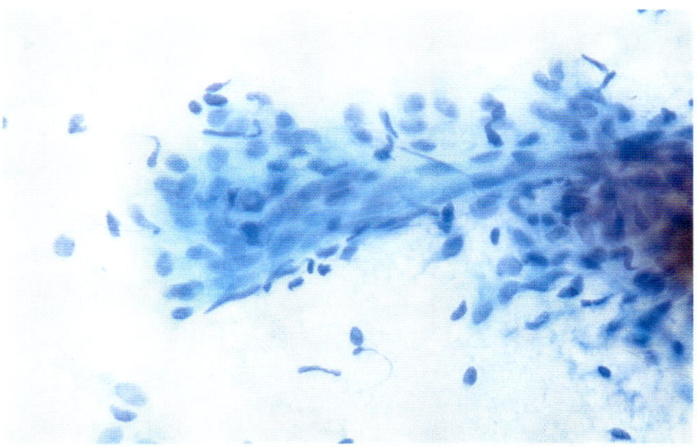

Figure 8.13 High-power view of loosely arranged stromal cells with pale basophilic cytoplasm and oval nuclei. Nuclear chromatin is even, nuclear outlines are regular and there are no nucleoli. ×400.

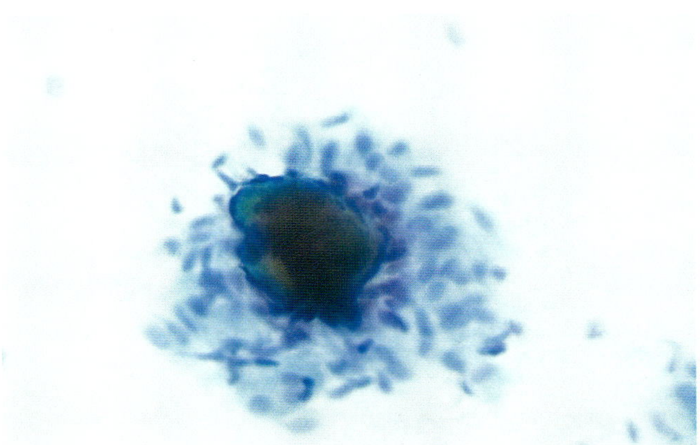

Figure 8.14 A solid psammomatoid mass of tissue enclosed by loosely cohesive, pale oval cells. These features, together with the clinical features, are compatible with a paranasal fibro-osseous lesion. ×400.

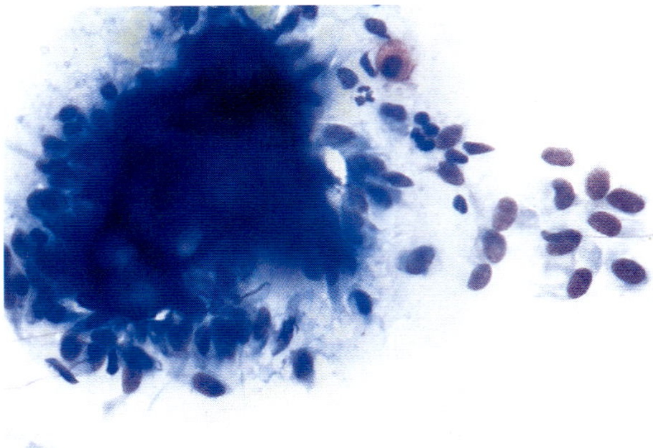

Figure 8.15 Air-dried intraoperative smear stained with Giemsa shows groups of tightly cohesive oval cells surrounding a solid mass. Loose-lying round to oval cells are lying in the background. They are similar to those seen in Figures 8.12 and 8.13. Giemsa. ×400.

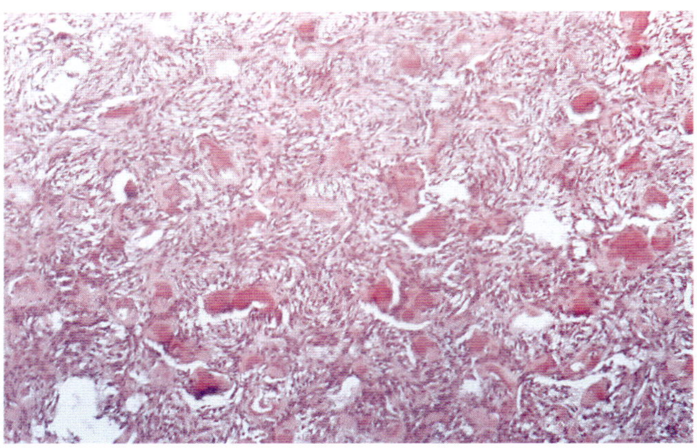

Figure 8.16 Histological section of the resected lesion shows a very cellular stroma enclosing islands of acellular psammomatoid bodies. These have been called small round ossicles (Margo *et al.*, 1995). These features are identical to those seen in Figures 8.9–8.13. Transition between ossifying fibroma and fibrous dysplasia can be discerned. H&E ×40.

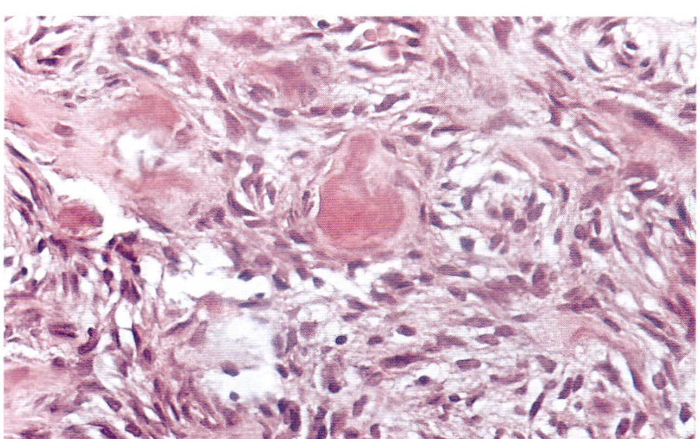

Figure 8.17 High-power view of the histological section shows cellular stroma and remodelling of ossicles with irregular cement lines. These features are those of a fibro-osseous lesion with psammomatoid features as described by Margo *et al.* (1995). ×400.

SINONASAL SQUAMOUS CARCINOMA

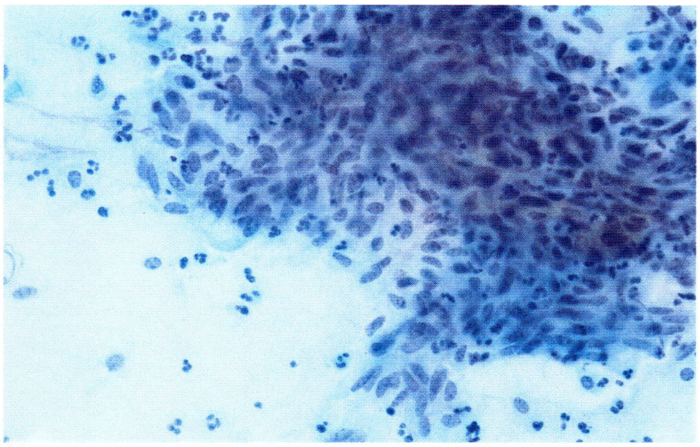

(a)

Figure 8.18 (a) This is a cytobrushing from a 21-year-old female with an 'inverted papilloma' of the lower lateral nasal wall, involving the inferior turbinate and inferior meatus. There are numerous loose-lying overlapping elongated cells with scanty cytoplasm. The nuclei are fairly uniform and the chromatin is disturbed. These cells resemble transitional epithelial cells. The possibility of a low-grade neoplasm is suggested by the cytological features. ×400.

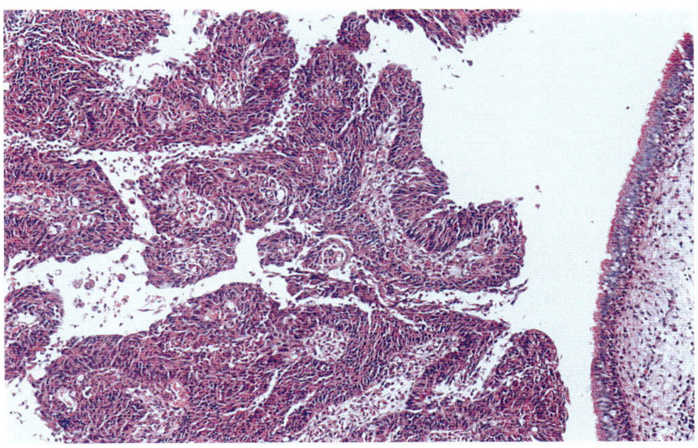

(b)

Figure 8.18 (b) This histological section of the nasal lesion shown in (a) shows an exophytic papillary-type growth pattern. Many of the cells have a transitional appearance and have a long axis perpendicular to the surface. There is cytological atypia. Foci suggestive of early stromal invasion are present. These features are consistent with a sinonasal squamous carcinoma of the lateral nasal wall.

NASAL RHABDOMYOSARCOMA

For a discussion of rhabdomyosarcoma in the head and neck see Chapter 9.

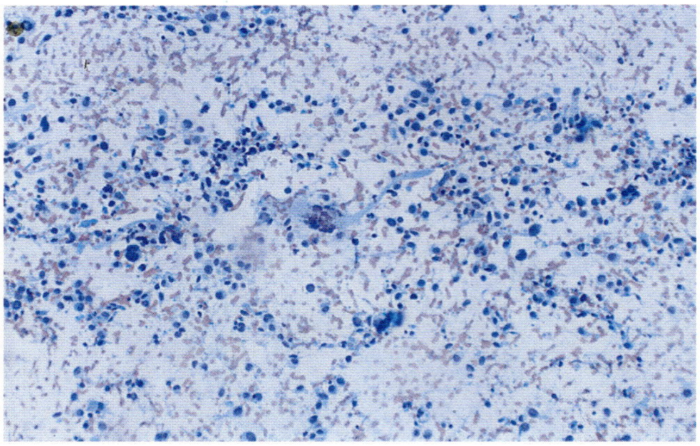

Figure 8.19 This cytobrushing of a haemorrhagic friable mass protruding from the nose in a 10-year-old boy shows a very cellular specimen. The cells are dissociated, small and round. At low power the pleomorphic nature of the cells can be seen. A huge multinucleated cell with a long cytoplasmic process is present. These appearances are more suggestive of a rhabdomyosarcoma, rather than a small round-cell ('blast' cell) tumour. ×100.

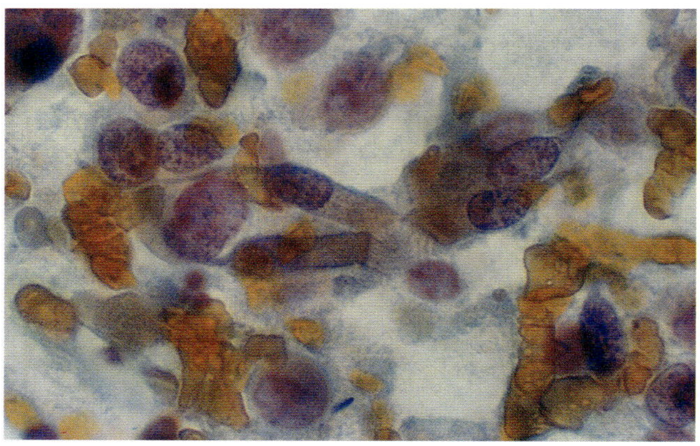

Figure 8.20 High-power view of the nasal brushing shown in Figure 8.19, showing reactive ciliated cells among large, round, undifferentiated, pleomorphic cells. The latter have scanty cytoplasm, rounded pleomorphic nuclei, disturbed nuclear chromatin and prominent irregular eosinophilic nucleoli. ×400.

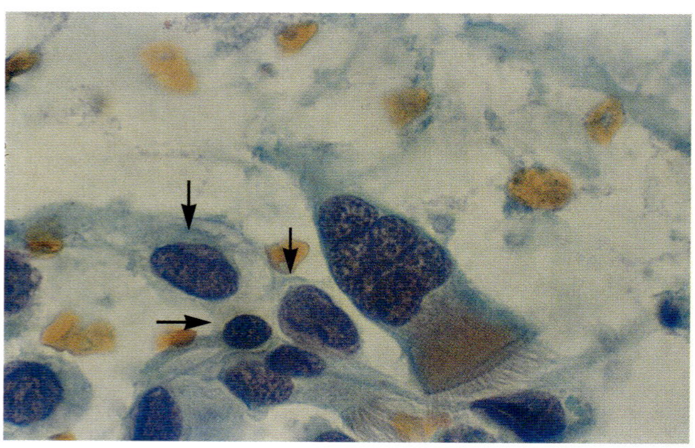

Figure 8.21 High-power view of the specimen shown in Figures 8.19 and 8.20, showing a *reactive* ciliated multinucleated respiratory cell with a well-defined terminal bar. The multiple nuclei show moulding and have markedly disturbed, speckled nuclear chromatin. No nucleoli are present. Ciliated cells are never malignant. A few less well-defined reactive ciliated cells are also present. Smaller, round, apparently undifferentiated cells with prominent nucleoli are also present (arrows). The latter have irregular nuclear outlines – a hallmark of malignancy. Markedly reactive changes are often seen in epithelial cells which are adjacent to a rapidly growing neoplasm. ×400.

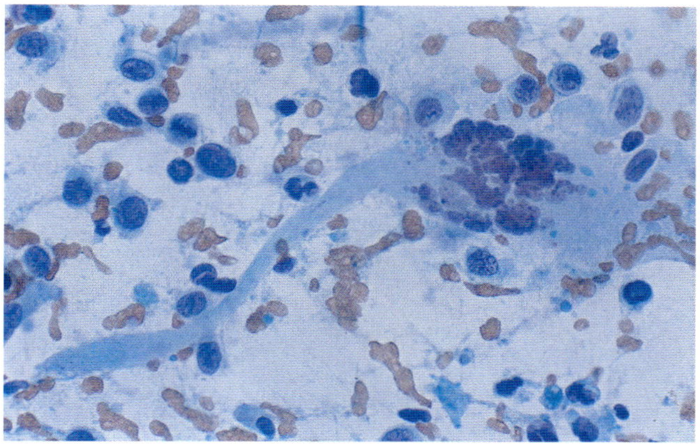

Figure 8.22 Same case as shown in Figures 8.19–8.21. High-power view of a multinucleated malignant cell with a long cytoplasmic process: a huge cell with a bizarre outline resembling a lacrosse racquet (hence the term 'racquet cell'). The plentiful cytoplasm contains multiple pleomorphic nuclei. Some of these show degenerative changes and apoptosis. Nucleoli are not obvious, but the nuclear chromatin shows clumping. These features are characteristic of a pleomorphic rhabdomyosarcoma of childhood. ×400.

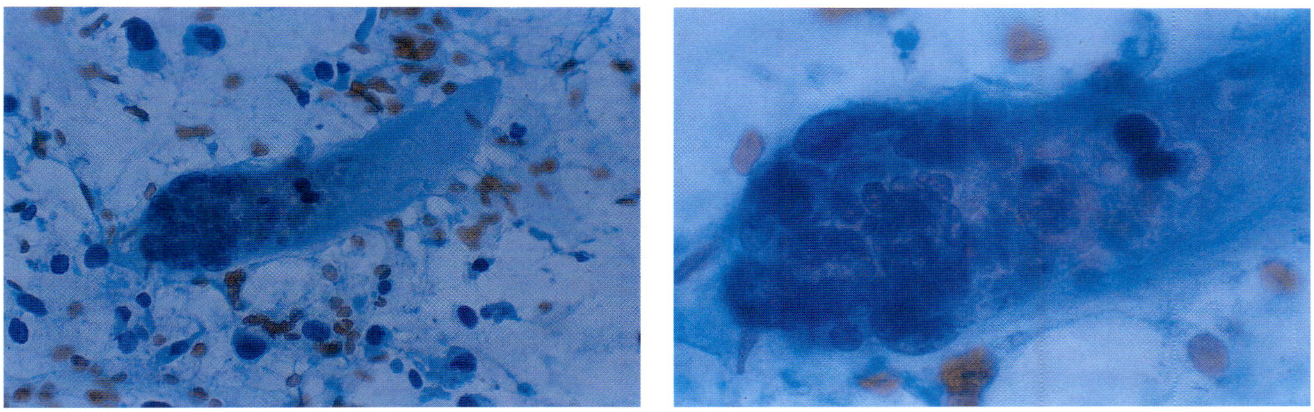

(a) (b)

Figure 8.23 (a) and (b) Another bizarre multinucleated cell from the same case as shown in Figure 8.22. Numerous smaller, pleomorphic hyperchromatic cells are present in the background. The oil-immersion view shows the crowded pleomorphic nuclei in greater detail. Some show early degenerative changes. Oil immersion. ×400 and ×1000.

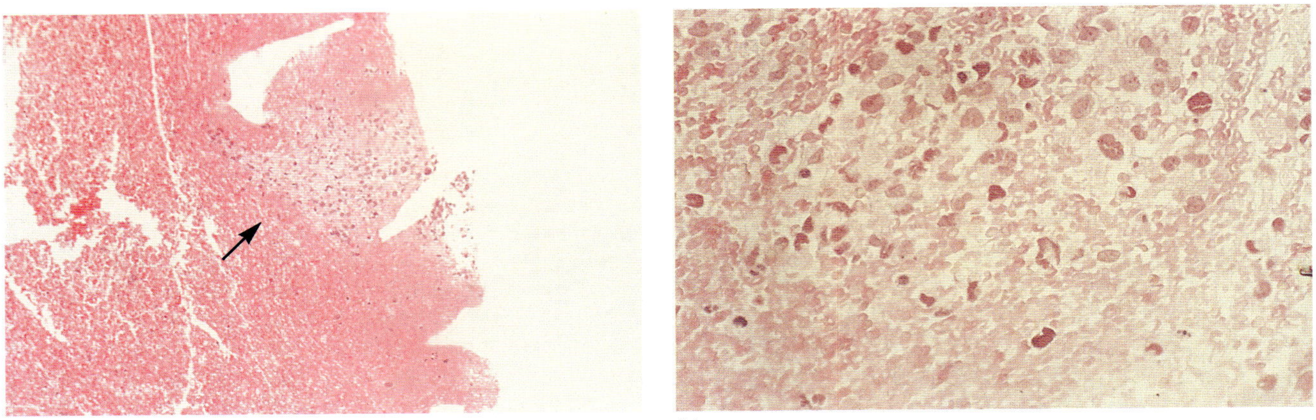

(a) (b)

Figure 8.24 (a) and (b) Attempted biopsy of the friable haemorrhagic nasal mass discussed in Figures 8.19–8.23 was difficult and yielded fragmented necrotic tissue. This was infiltrated by small degenerate undifferentiated cells lying in fresh blood. They had no distinguishing features. ×100 and ×400.

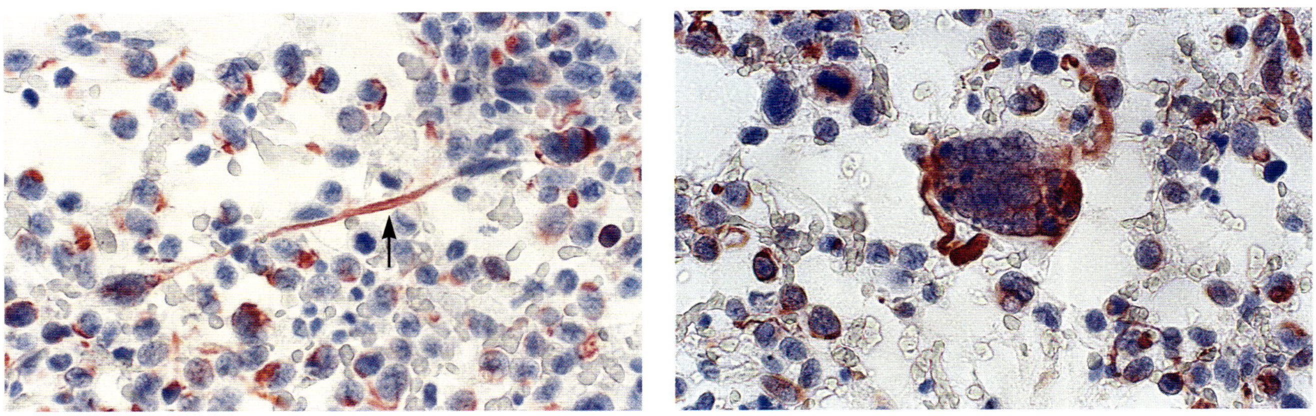

(a) (b)

Figure 8.25 (a) and (b) Owing to the unsatisfactory nature of the biopsy of the nasal mass discussed in Figures 8.19–8.24, one of the cytobrush slides was destained and then stained with an immunocytochemical stain for desmin (Michaels, 1990). This is strikingly positive, and highlights not only the sarcomatoid origin of the cells but their pleomorphism as well. Possible cross-striations are seen in the cytoplasm of the elongated cell. The scanty rim of cytoplasm in the small 'undifferentiated' cells is also positive for desmin. The diagnosis of rhabdomyosarcoma (mixed embryonal and pleomorphic) was accepted (see Chapter 9). ×100 and ×400.

LIPOMA OF THE PARANASAL/NASOPHARYNGEAL AREA

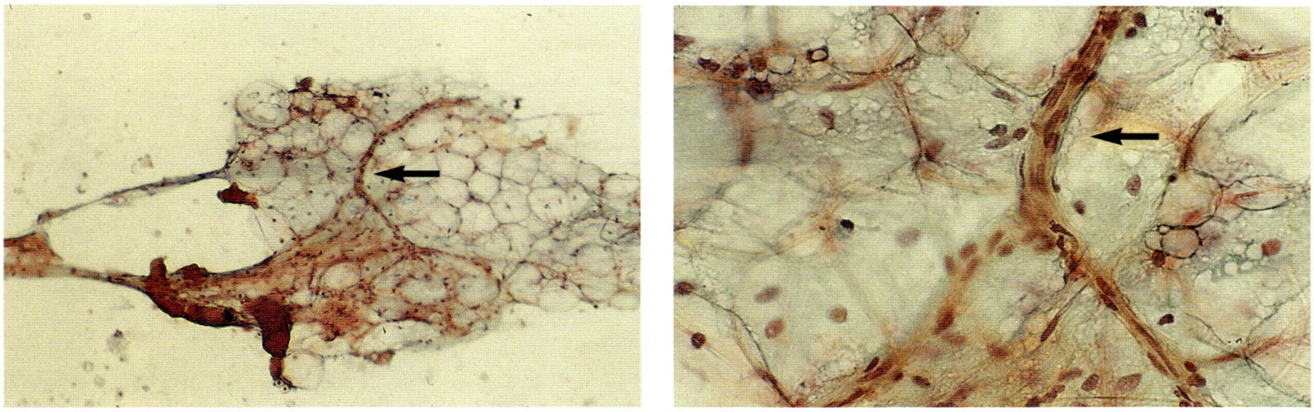

(a) (b)

Figure 8.27 (a) and (b) FNA specimen of a paranasal mass which appeared as a soft-tissue mass/cystic mass on tomograms in a patient complaining of nasal obstruction. The aspirate consisted of sheets of adipose tissue with prominent blood vessels coursing through it. The high-power view shows a blood vessel containing longitudinal smooth muscle cells in the wall. These appearances are compatible with a lipoma, though the term angiolipoma could also be applied. These benign neoplasms are rare in the head and neck (Fagan *et al*., 1996; Fu and Perzin, 1977). ×100 and ×400.

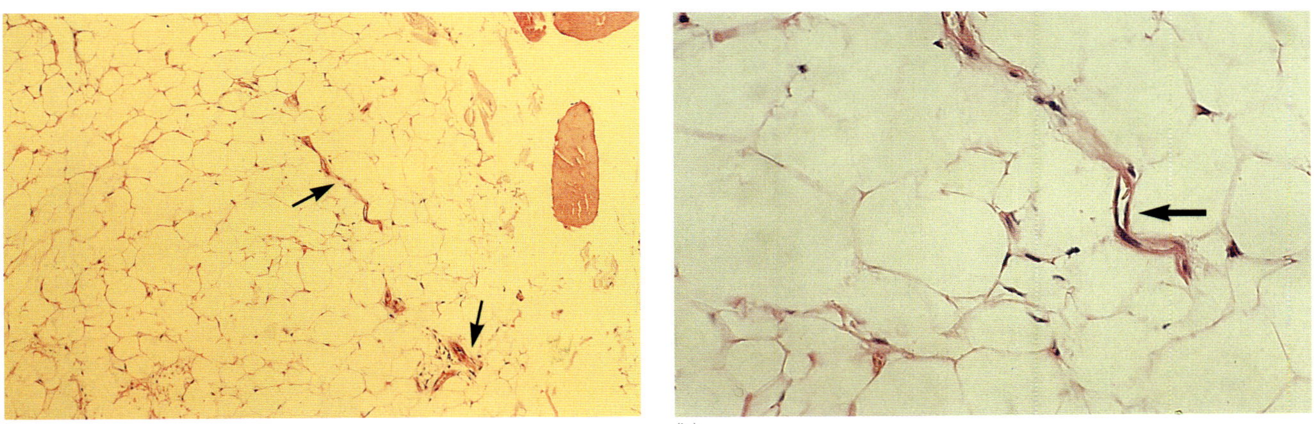

(a) (b)

Figure 8.28 (a) and (b) Histological section of the lesion shown in Figure 8.27 shows adipose tissue incorporating skeletal muscle fibres as well as prominent arterioles (arrow). The diagnosis of lipoma/angiolipoma was accepted. H&E ×40 and ×100.

REFERENCES AND FURTHER READING

General

Das, D. K., Gulati, A., Bhatt, N. C. *et al.* (1993) Fine-needle aspiration of oral and pharyngeal lesions. *Acta Cytol.* **37**: 333–342.

Perzin, K. H. and Fu, Y. S. (1984) Non-epithelial tumours of the nasal cavity, paranasal sinuses and nasopharynx: a clinicopathological study XIII. *Cancer* **54**: 1860–1869.

Warner-Learmonth, G. M., Duff, J. and Sellars, S. (1993) The cytological diagnosis of nasopharyngeal neoplasms using the flexible Endobrush. *Pathol. Res. Pract.* **189**: 840.

Nasal rhabdomyosarcoma

Ahkter, M., Ali, M. A., Bakry, M., Hug, M. and Sackey, K. (1992) Fine-needle aspiration diagnosis of rhabdomyosarcoma: cytologic, histological and ultrasound correlations. *Diagn. Cytopathol.* **8**: 465–474.

Chetty, R., Learmonth, G. M., Price, S. and Taylor, D. (1991) Primary oesophageal rhabdomyosarcoma. *Cytopathology* **2**: 103–108.

Hellquist, H. B. (1990) *Pathology of the Nose and Paranasal Sinuses.* London, Butterworths, p. 131.

Howell, L. P. (1992) The cytology of paediatric masses: a differential diagnostic approach. *Diagn. Cytol.* **8**: 107–115.

Michaels, L. (1990) *Atlas of Ear, Nose and Throat Pathology.* London, Kluwer, p. 69.

Travis, W. D. and Wold, L. E. (1987) Immunoperoxidase staining of fine-needle aspiration specimens previously stained by the Papanicolaou technique. *Acta Cytol.* **31**: 517–520.

Fibro-osseous lesion

Margo, C. E., Ragsdale, B. D., Perman, K. I., Zimmerman, L. E. and Sweet, D. E. (1995) Psammomatoid (juvenile) ossifying fibroma of the orbit. *Ophthalmology* **92**: 150–159.

McCullough, J. B., Evans, A.T., Vaughan-Jones, R. and Hussein, K. A. (1996) Fine-needle aspiration of a nasal meningioma: a case report. *Cytopathology* **7**: 56–61.

Michaels, L. (1990) *Atlas of Ear, Nose and Throat Pathology.* London, Kluwer, p. 61.

Nasopharyngeal lipoma

Fagan, J. J., Learmonth, G. M., Garb, M. and Bowen, R. M. (1996) Nasopharyngeal lipoma: a rare clinicopathological entity. *J. Laryngol. Otol.* **110**: 275–276.

Fu, Y. S. and Perzin, K. H. (1977) Non-epithelial tumors of the nasal cavity, paranasal sinuses and nasopharynx. VIII Adipose tissue tumors (lipoma and liposarcomas). *Cancer* **40**: 1314–1317.

9 Skin and soft tissue

Fine-needle aspiration of the skin and soft tissues of the neck is useful in confirming neoplastic recurrences and metastases, and in determining the nature of non-malignant lesions of soft tissue. FNA of primary neoplasms of skin has been attempted; however, in the author's opinion it is technically difficult owing to the superficial nature of the lesions. Furthermore, it is of little clinical value as most skin lesions are excised *in toto* for histological examination, with assessment of the peripheral and deep lines of excision.

SUPERFICIAL SKIN LESIONS

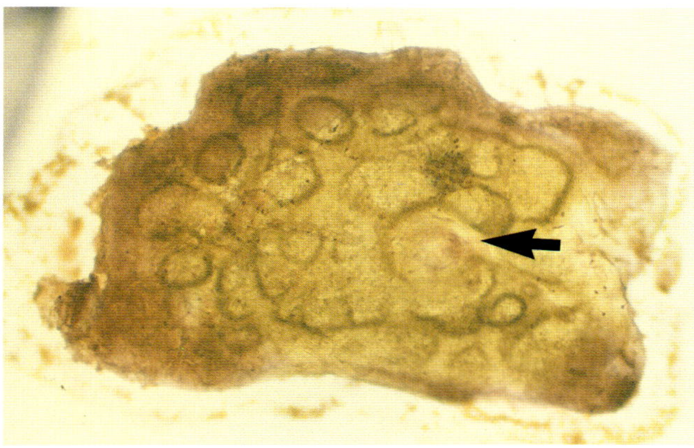

Figure 9.1 FNA of a nodule in a parotidectomy scar from a patient who had had a pleomorphic adenoma excised many years previously. Clinically a carcinoma was suspected. A large intact microbiopsy of pigmented tissue is shown (the patient is of the Xhosa race). It has a three-dimensional appearance. The pigmented cells show regular peripheral palisading around groups of cohesive orderly non-pigmented cells. Note the keratin plug (arrow). No malignant epithelial cells were present. These appearances are suggestive of normal pigmented skin with pigmented melanocytes surrounding the rete ridges. This suggests that the mass in the scar was possibly implanted skin (implantation dermoid). This case illustrates the importance of clinicopathological correlation, including the racial group of the patient. If the patient was Caucasian, a differential diagnosis of a pigmented basal cell carcinoma would have been offered. ×40.

SUPERFICIAL SQUAMOUS CARCINOMA OF SKIN

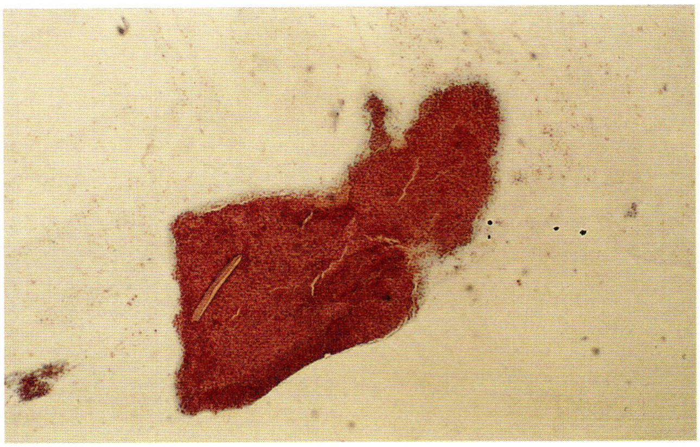

(a)

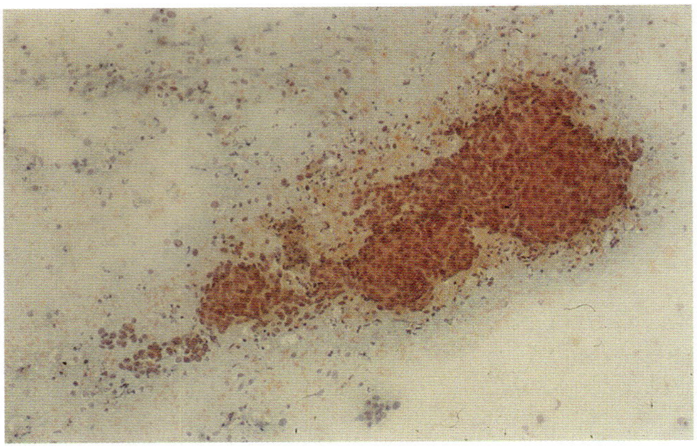

(b)

Figure 9.2 (a) and (b) Cytobrushing of a persistent ulcer in the ear canal of an elderly farmer of Celtic descent. Sheets of small, crowded but cohesive hyperchromatic cells are present. Some of the loose-lying cells show moderate nuclear pleomorphism. These cytological appearances are similar to those seen in severe dysplasia/carcinoma in situ on cervical smears. ×40. ×100.

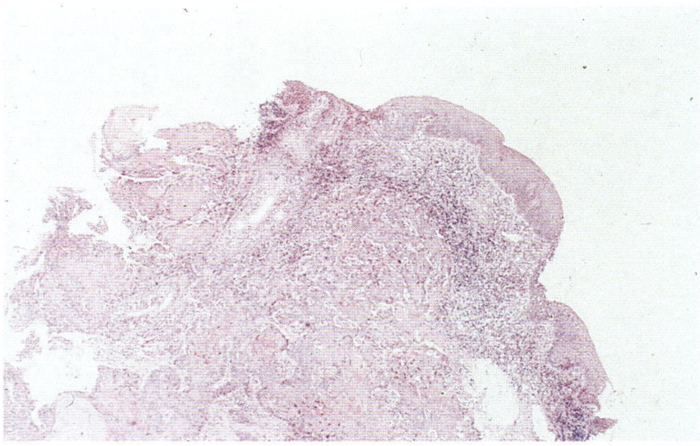

(c)

Figure 9.2 (c) A biopsy, though clinically difficult, was attempted; this showed invasive squamous carcinoma, well differentiated. H&E ×40.

BASAL CELL CARCINOMA OF SKIN

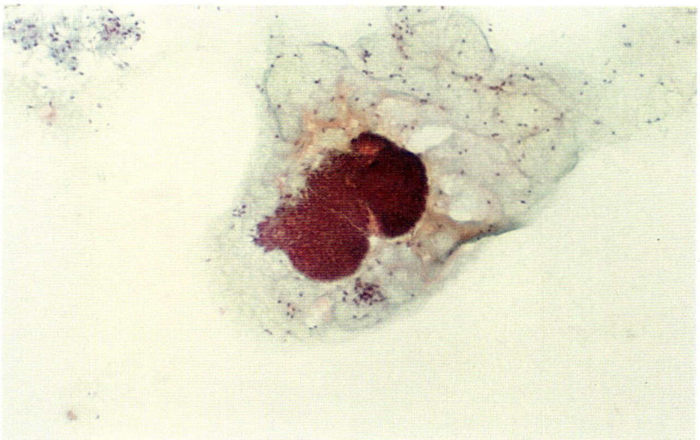

(a)

Figure 9.3 (a) FNA of a superficial nodule on facial skin overlying the maxillary ramus which was clinically a 'rodent ulcer'. A cohesive microbiopsy of tissue with rounded contours, composed of small, uniform basophilic cells is present. It appears to be associated with sebaceous tissue. No melanin pigment is present. These features are compatible with a basal cell carcinoma. ×40.

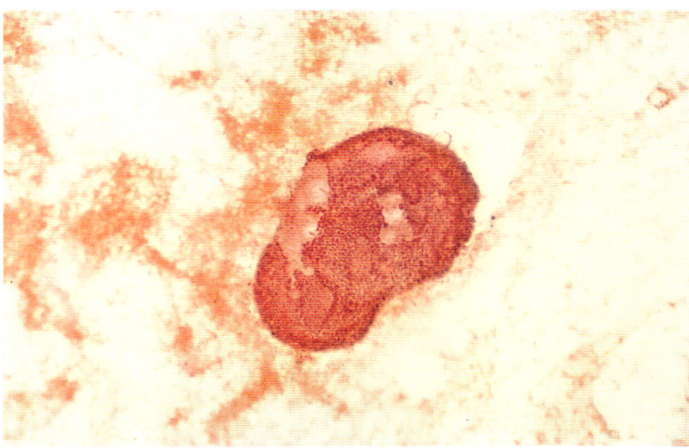

(b)

Figure 9.3 (b) A microbiopsy of aspirated basal cell carcinoma showing cohesive clusters of small uniform basophilic cells with apparent cribriform arrangement. A biopsy was technically impossible. ×40.

SQUAMOUS CARCINOMA OF THE NASAL COLUMNELLA

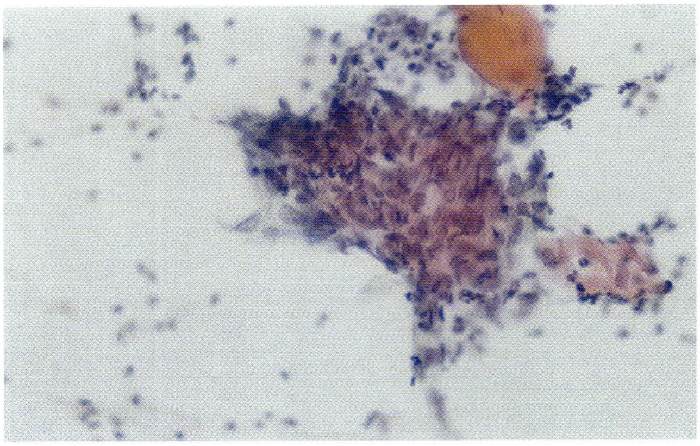

Figure 9.4 (a) Cytobrushing of long-standing irritated lesion of the nasal columnella in an elderly Caucasian patient. A sheet of small, crowded, highly atypical epithelial cells and a mass of keratinous tissue is shown. These features are compatible with severe dysplasia, at least. An invasive lesion is suspected. ×100.

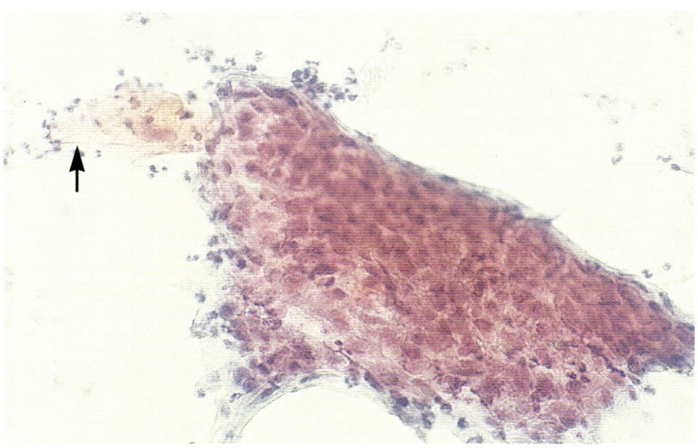

Figure 9.4 (b) Cytobrushing of an ulcerating lesion in the nasal columnella of an elderly patient which was technically too difficult to biopsy. Clinically there was an invasive lesion. A 'microbiopsy' or sheet of crowded moderately pleomorphic epithelial cells is shown. The cells are disorganized and show all the features of severe squamous dysplasia at least. A normal angular eosinophilic squamous cell is present (arrow). ×100.

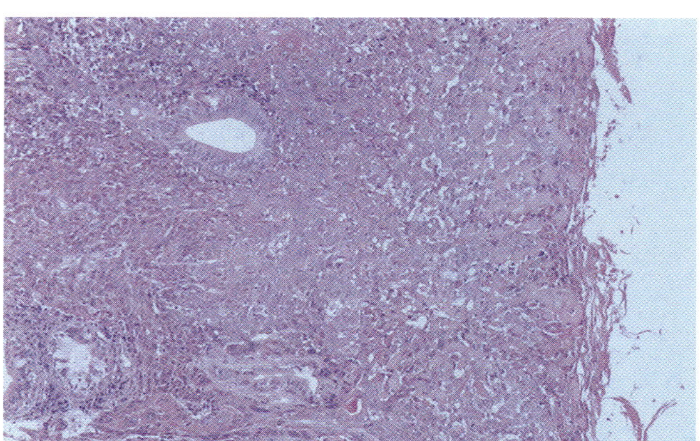

Figure 9.4 (c) Histological biopsy of the lesion shown in (a) shows an invasive keratinizing squamous carcinoma of the nasal columnella. ×400.

RECURRENT CARCINOMA IN A SCAR

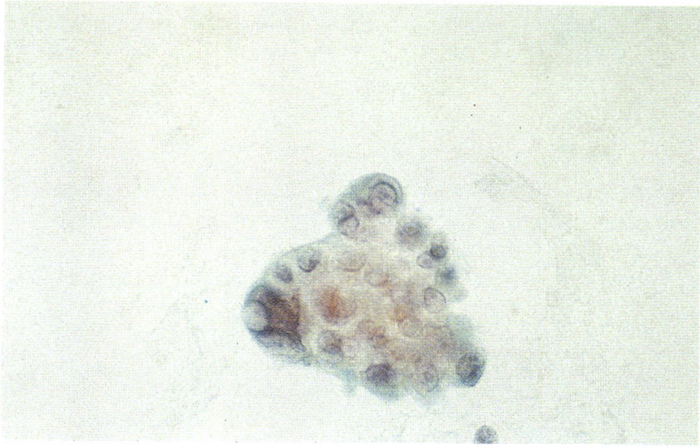

(a)

Figure 9.5 (a) FNA of a parotidectomy scar. The histogenesis of the previously excised parotid neoplasm was unknown at the time of aspiration. A group of pleomorphic malignant cells with large nuclei and prominent nucleoli is shown. Although the contour of the cell group is rounded, there is no three-dimensional effect. A large pleomorphic cell on the edge of the cell group is vacuolated. These features are suggestive of mixed squamoid and glandular origin. ×400.

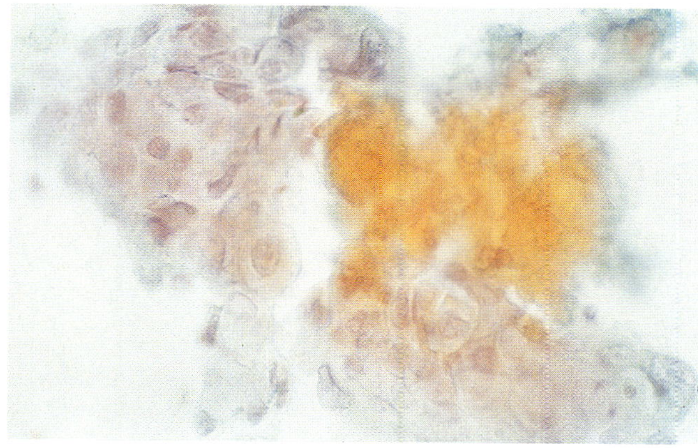

(b)

Figure 9.5 (b) Another group of cells from the same specimen shows pleomorphic malignant cells intermixed with eosinophilic material resembling keratin. These appearances are compatible with a mucoepidermoid carcinoma. ×400.

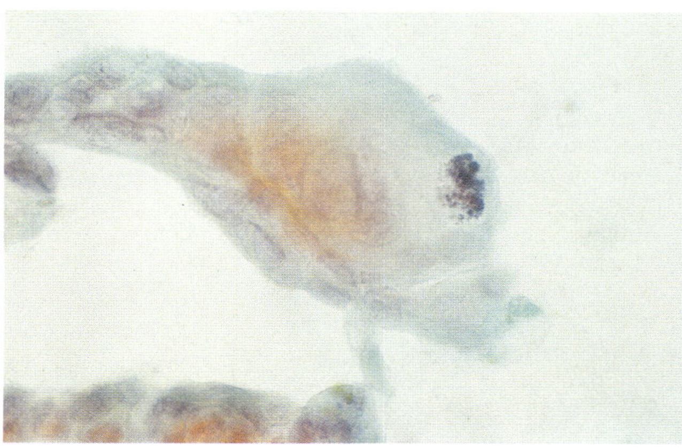

(c)

Figure 9.5 (c) Another view, showing a large bizarre anaplastic cell with squamoid features, multiple pleomorphic nuclei and prominent eosinophilic nucleoli. Some degenerate nuclear material is present in the cytoplasm. ×400.

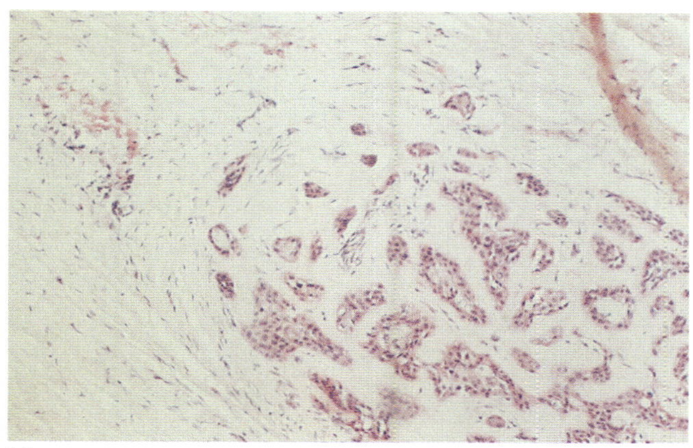

(d)

Figure 9.5 (d) Trucut biopsy of the mass in the scar showed a recurrent mucoepidermoid carcinoma. Further investigation confirmed that the original neoplasm was a mucoepidermoid carcinoma. H&E ×40.

Mucoepidermoid carcinoma is one of the most difficult salivary gland tumours to categorize accurately on cytology (MacLeod and Frable, 1993); however, for the purposes of clinical management it usually suffices to tell the clinician that the lesion is malignant and could possibly be a mucoepidermoid carcinoma.

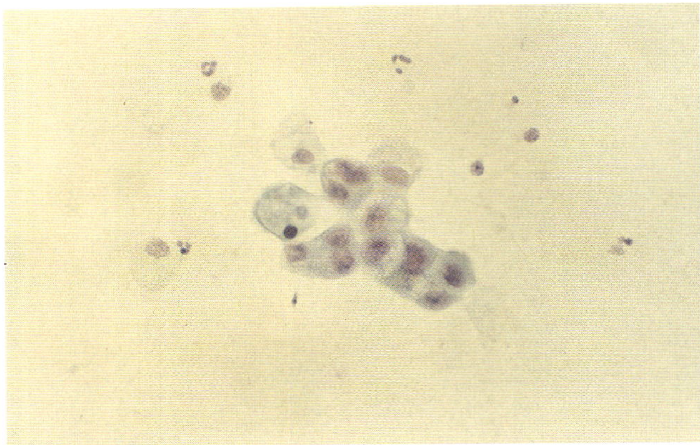

(a)

Figure 9.6 (a) FNA of a superficial skin nodule in a patient with gastric carcinoma. A three-dimensional cluster of pleomorphic cells is present. They have vacuolated clear cytoplasm, well-defined cytoplasmic borders and large nuclei with prominent nucleoli. These appearances are compatible with metastatic gastric carcinoma. ×400.

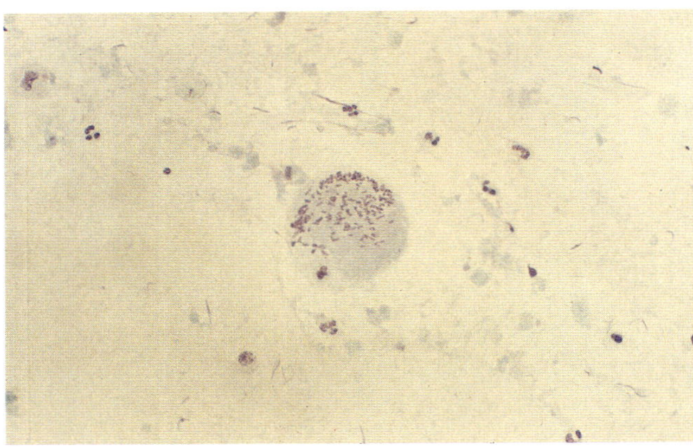

(b)

Figure 9.6 (b) Same case as shown in (a): a huge malignant cell with an abnormal mitotic figure. Individual chromosomes can be discerned. ×400.

HERPES VIRUS INFECTION

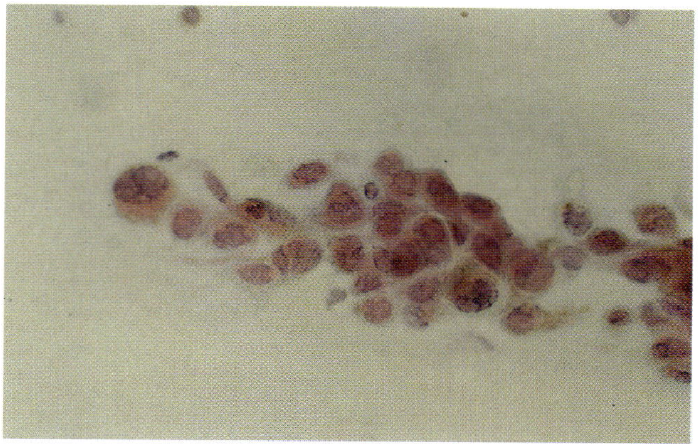

Figure 9.7 Cytobrush specimen of a cutaneous vesicle close to the upper lip. The flat sheet of multinucleated cells shows early degenerative changes. The nuclei have a 'ground-glass' appearance, with pale clumped chromatin, suggestive of herpes infection. Some of the nuclei have angular outlines and have been likened to 'crazy-paving'. Some of the cells are binucleate. It is not possible to differentiate herpes zoster from herpes simplex infection on cytology (Grubb, 1988). ×400.

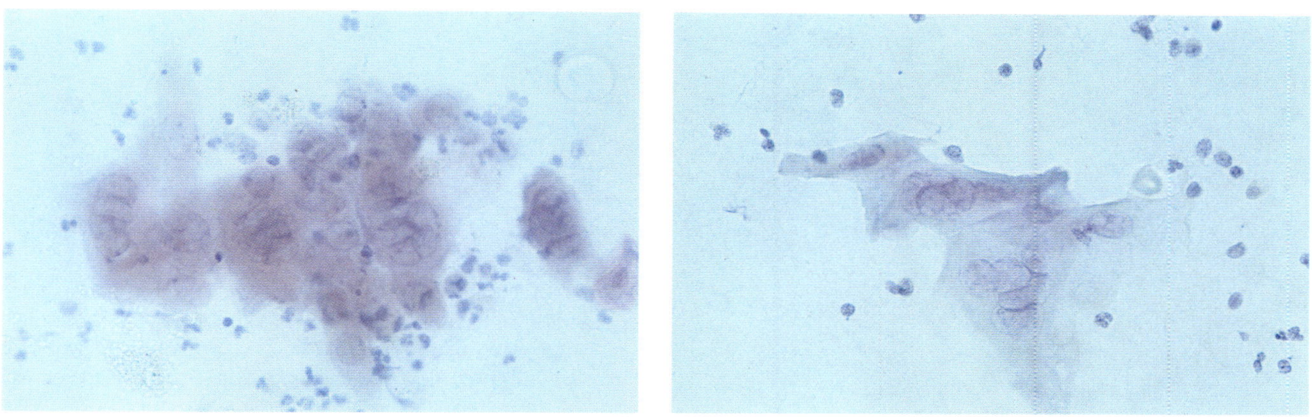

(a) (b)

Figure 9.8 (a) and (b) Cytobrush specimen from a neck wound in an HIV-positive patient. Large multinucleated cells are present in a background of mixed inflammatory cells. The nuclei show characteristic moulding or 'crazy-paving'. The nuclear chromatin is very pale. These features are suggestive of herpes infection. ×400.

Cellular material from superficial ulcers or vesicles usually shows marked degenerative changes with pale nuclear staining. Cytobrush specimens yield superior material to swabs or scrapes. The cytobrush may also be rinsed in a suitable fluid for bacteriological studies. Dense intranuclear inclusions may be seen in well-preserved cells; these are absent in more degenerate cells.

LIPOMA OF SOFT TISSUE OF THE NECK

Lipomas are usually palpable as soft, mobile symptomless masses. Local trauma may draw attention to them if fat necrosis occurs, resulting in tenderness and a change in consistency. Lipomas of the head and neck are considered rare (see Chapter 8) (Fu and Perzin, 1977; Fagan *et al.*, 1996).

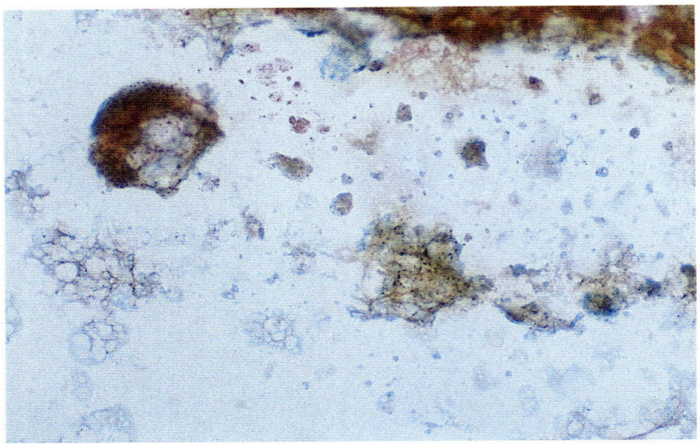

(a)

Figure 9.9 (a) FNA of a firm mass in the neck. Clusters of benign adipocytes are present, some showing loss of cohesion and cloudiness suggestive of fat necrosis. No epithelial cells are seen. ×100.

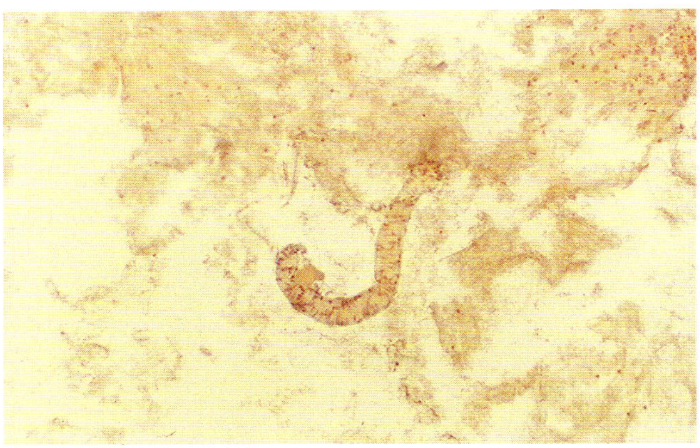

(b)

Figure 9.9 (b) Photomicrograph of another area of the specimen discussed in (a), showing a blood vessel (probably an arteriole) in a background of amorphous debris. Note the circular and longitudinal orientation of the smooth muscle cells of the arteriolar wall. No epithelial cells are present. ×100.

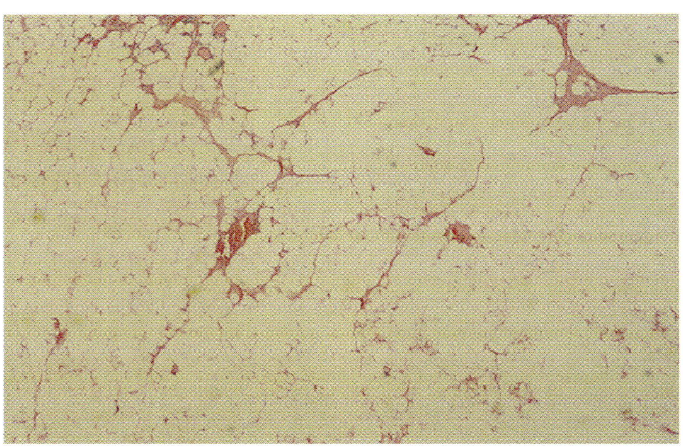

(c)

Figure 9.9 (c) Histological section of the same case showing a lipoma traversed by capillaries. H&E ×40.

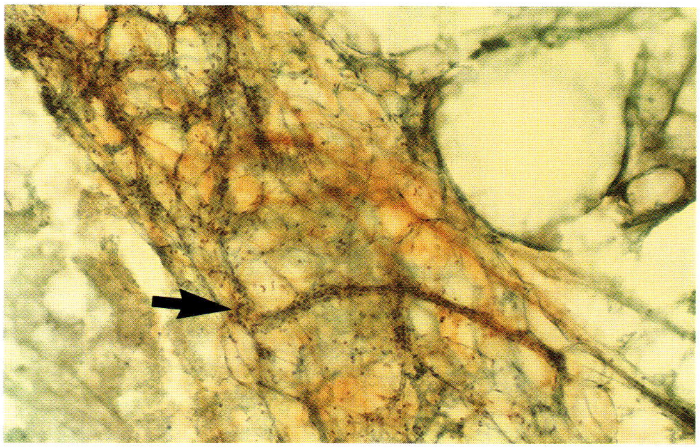

Figure 9.10 FNA of a firm, *painful*, fairly vascular subcutaneous mass in the neck, showing a sheet of benign adipose tissue traversed by blood vessels (arrow). These features, together with the clinical presentation, are suggestive of an angiolipoma. The latter are painful because of the presence of neural fibres in the vascular bundles. ×100. (Case courtesy of Dr M. du Preez, Paarl, South Africa.)

FIBROMYXOMA OF SOFT TISSUE

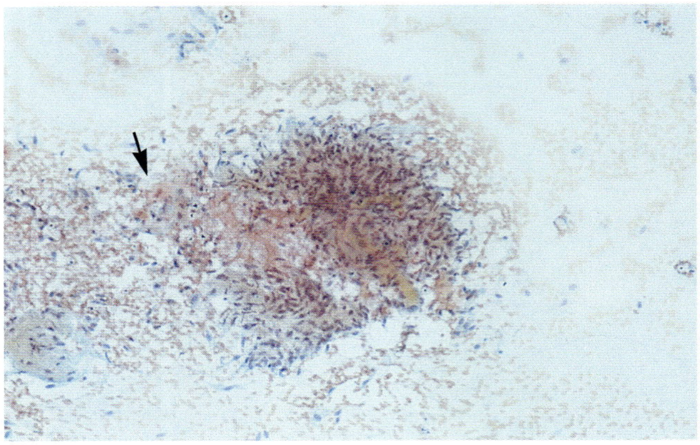

(a)

Figure 9.11 Fibromyxoma of the cheek. (a) This FNA from a young adult with a soft, fairly mobile mass in the cheek was obtained through the oral route (mucosa). This photomicrograph shows a loose collection of fairly uniform, pale-staining spindle-shaped cells with tails of cytoplasm. These are more suggestive of a soft tissue neoplasm than a malignant squamous lesion. Note the benign superficial squamous cells from the oral cavity (arrow). ×100.

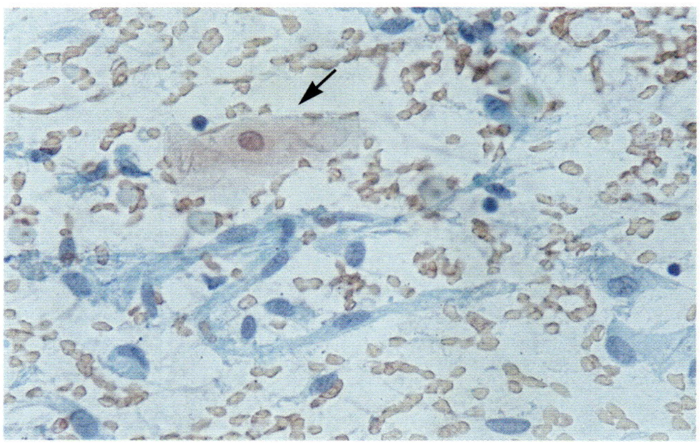

(b)

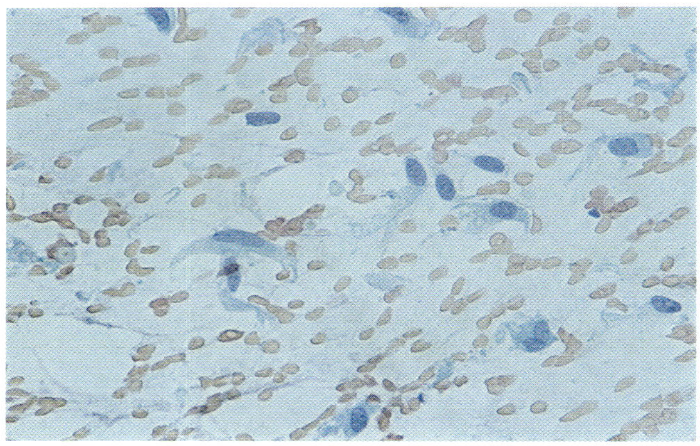

(c)

Figure 9.11 (b) and (c) High-power view of the lesion shown in (a) shows the loose-lying spindle cells with pale oval nuclei and long cytoplasmic tails. The nuclear outlines are regular. Tiny pinpoint nucleoli are seen. These features are compatible with a benign stromal neoplasm in a young adult. Note the benign oral squamous cell (arrow). ×400.

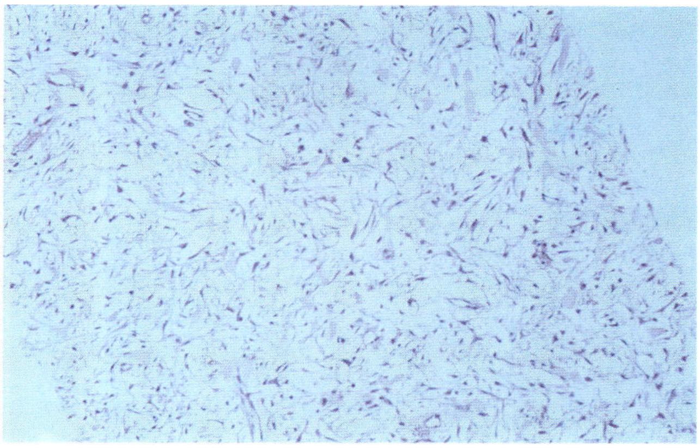

(d)

Figure 9.11 (d) Histological section of the mass shown in (a)–(c) shows features of a benign fibromyxoma. Note the loose-lying elongated pale cells in a pale stroma. Some of the cells have stellate features. H&E ×100.

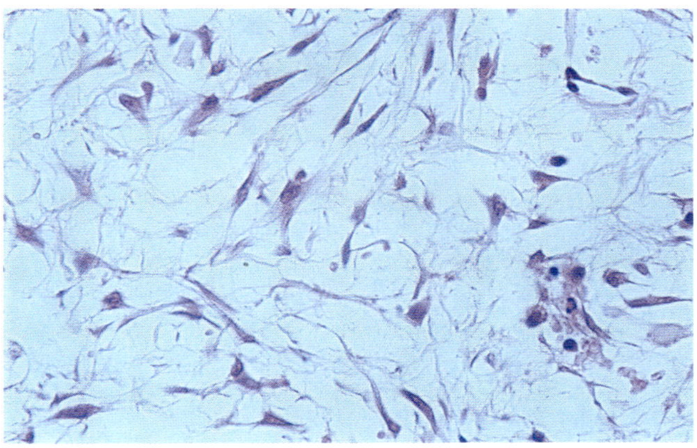

(e)

Figure 9.11 (e) High-power view of the fibromyxoma shown in (a)–(d), showing the elongated nature of the cells lying in a loose stroma. The stellate nature of the cells is more apparent than on the cytological smears. H&E ×400.

METASTATIC MALIGNANT LESIONS IN SOFT TISSUE OF THE NECK

Metastatic carcinomas usually spread to lymph nodes in the neck, rather than into the soft tissues. However, stromal neoplasms and malignant melanoma may spread directly to soft tissues in the neck. Clinically, if the lesion is diffuse it may be difficult to decide whether it is in lymph nodes or in the soft tissue. Clinical correlation is essential, if the correct advice with regard to further investigation is to be given to the clinician, in order to triage the patient correctly. It is helpful to comment on the presence or absence of lymphoid cells when reporting on metastatic lesions in the neck.

METASTATIC MALIGNANT MELANOMA IN SOFT TISSUE

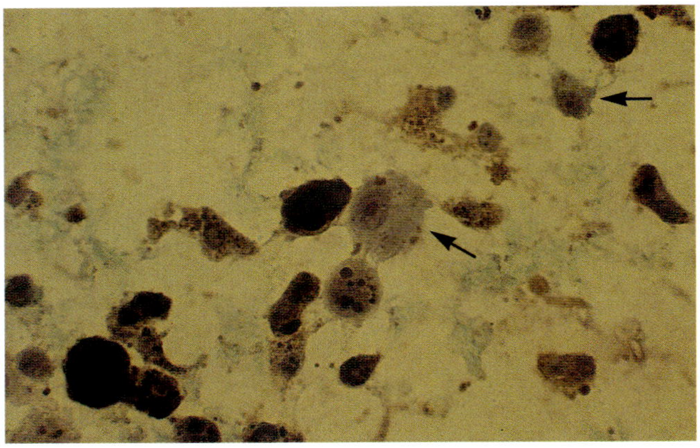

(a)

Figure 9.12 (a) FNA of a very large painless mass in the neck of a 28-year-old white male. Numerous heavily pigmented loose-lying cells are shown. The pigment is brown-black and does not appear to be refractile. It obscures much of the cellular detail. Many of the cells are possibly heavily laden macrophages; however, several large cells have plentiful cytoplasm, a large nucleus and a large prominent eosinophilic nucleolus (arrows). These raise the possibility of a metastatic malignant melanoma. The problem of pigments in FNA specimens of the head and neck is discussed in Chapter 4. ×400.

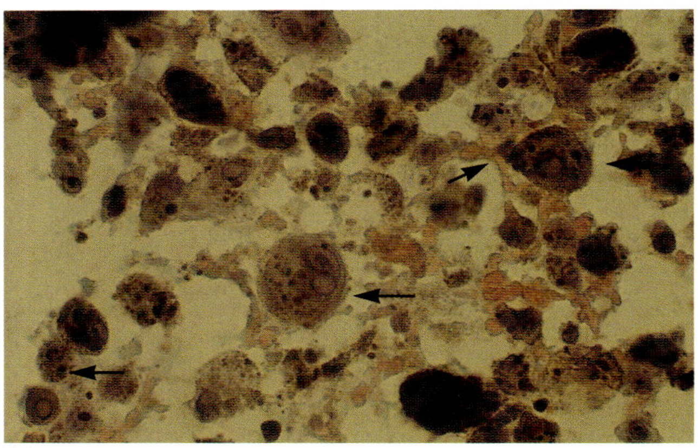

(b)

Figure 9.12 (b) Another view of the case illustrated in (a), showing innumerable pigment-laden macrophages. The pigment almost obscures the cell details; however, large cells with large nuclei and prominent nucleoli can be seen. An intracytoplasmic vacuole is noted in this cell and others (arrows). There was no history of previous excision of a pigmented lesion and no obvious pigmented lesions were noted on the skin. However, on consultation with a dermatologist a melanoma was found on the scalp, obscured by hair. For a discussion on the differential diagnosis of metastatic malignant melanoma in FNA specimens from the neck see Chapter 2.

METASTATIC CHONDROSARCOMA

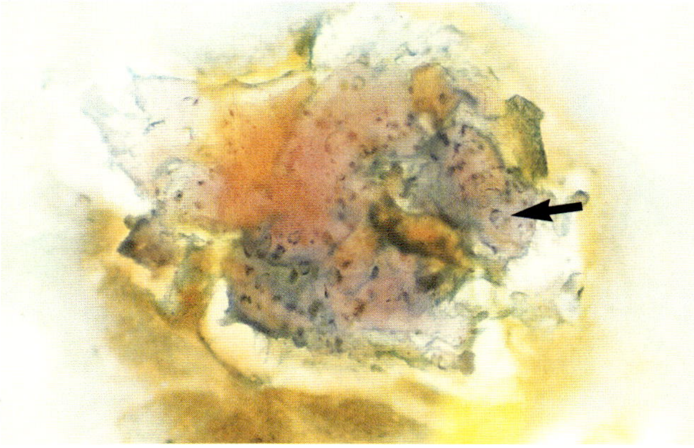

(a)

Figure 9.13 (a) FNA of a mass in the neck of a middle-aged male who also complained of pain around the shoulder joint. A microbiopsy of metachromatic tissue is shown. It has a lobular outline and is moderately cellular. Lacunar cells are seen (arrow). No loose-lying cells or epithelial cells are seen. The appearances suggest a soft-tissue neoplasm, possibly myxoid or chondroid. Further investigation of the patient revealed a lytic lesion around the shoulder joint. ×100.

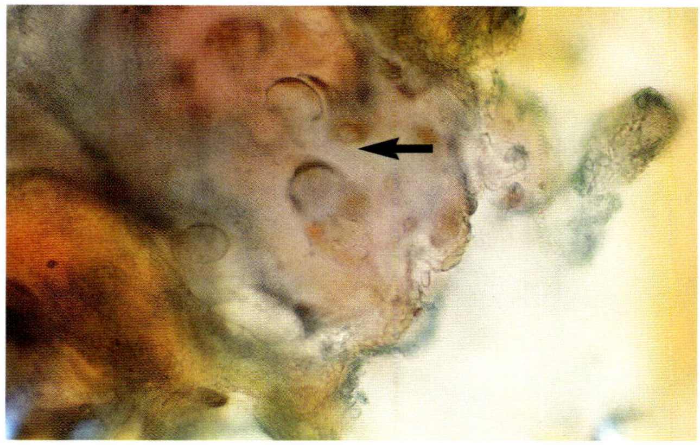

(b)

Figure 9.13 (b) High-power view of the mass discussed in (a), showing the characteristic metachromatic appearance of a chondroid lesion, with typical chondrocytic lacunar cells. The latter appear to have a halo around the nucleus. ×400.

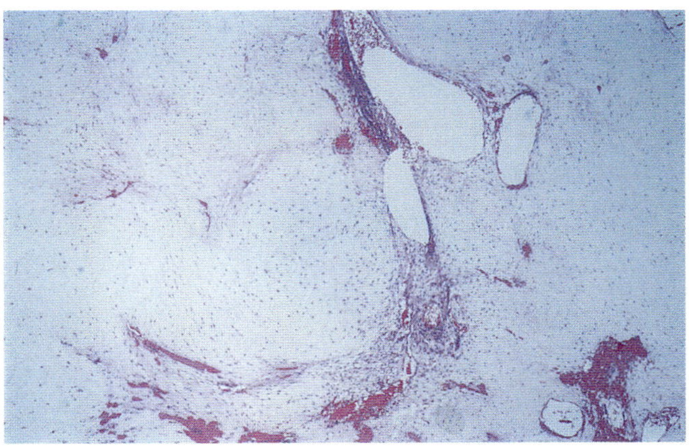

(c)

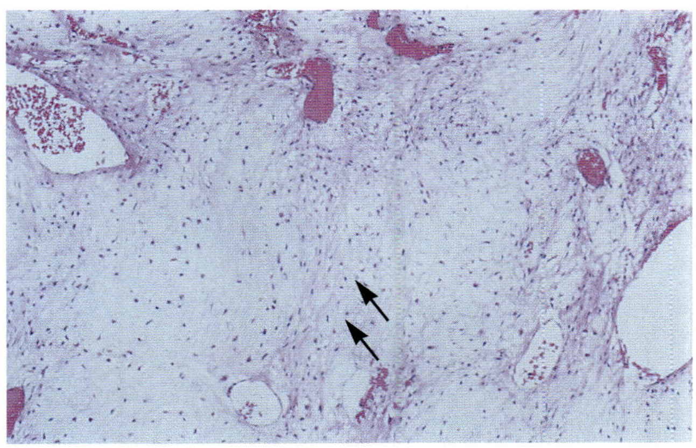

(d)

Figure 9.13 (c) and (d) Histological section of the lesion showed a chondrosarcoma of intermediate grade. On H&E stain the stroma appears very pale – almost translucent, with a lobulated pattern. The chondrocytic lacunar cells (arrow) are moderate in number; they are well demarcated and quite characteristic. H&E ×40 and ×100.

It is not possible to grade chondroid neoplasms on cytology (Stormby and Akerman, 1973; Sanerkin and Jeffree, 1990). It suffices to inform the clinician that it is a chondroid lesion and that further investigation is necessary. As a rule of thumb it is safe to say that if there are numerous loose-lying cells and very cellular microbiopsies it is probably a high-grade lesion, whereas if the microbiopsies present are sparsely cellular it is probably low grade.

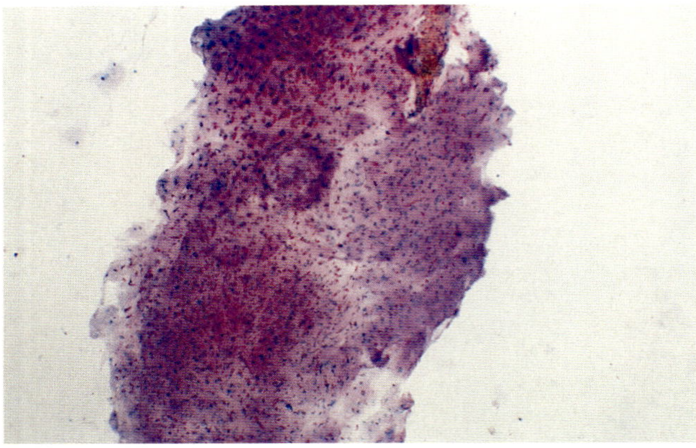

(a)

Figure 9.14 (a) FNA of a large mass in the neck, thought clinically to be a mass of matted lymph nodes. The patient, aged 50, also complained of pain in the region of the scapula. A microbiopsy of quite cellular, magenta-coloured metachromatic material is shown. A lobular outline is apparent. The possibility of a myxoid or chondroid lesion should be considered. ×100.

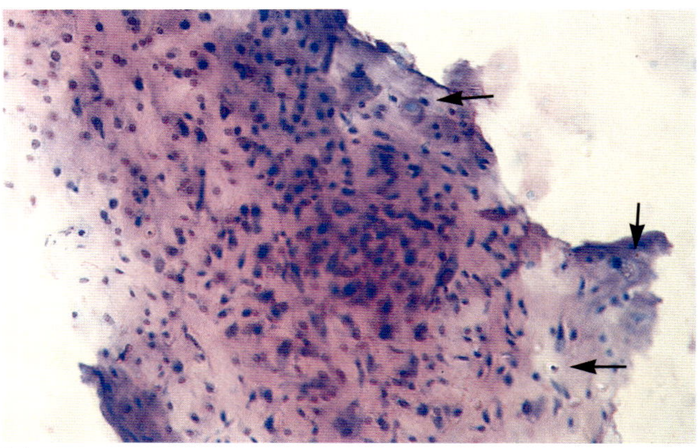

(b)

Figure 9.14 (b) High-power view of the lesion shown in (a), showing the markedly cellular metachromatic tissue. A few lacunar cells are obvious (arrows). Further investigation revealed a high-grade chondrosarcoma of the scapula. ×400.

NEUROFIBROMA V. NEUROFIBROSARCOMA

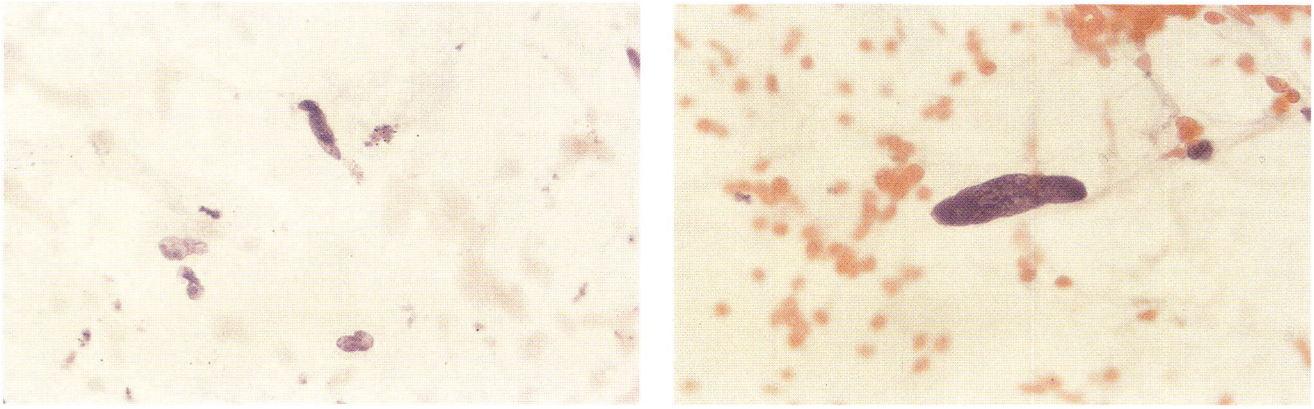

(a) (b)

Figure 9.15 (a) and (b) FNA of a soft tissue mass in the neck of a middle-aged woman. Single-lying, oval and spindle-shaped pleomorphic nuclei are shown in a clear background. Nuclear outlines are irregular. The nuclear chromatin has a speckled ('salt-and-pepper') appearance. No necrosis, blood or myxoid material is seen. Sarcomatous cells generally show speckled chromatin on FNA specimens. ×100. ×400.

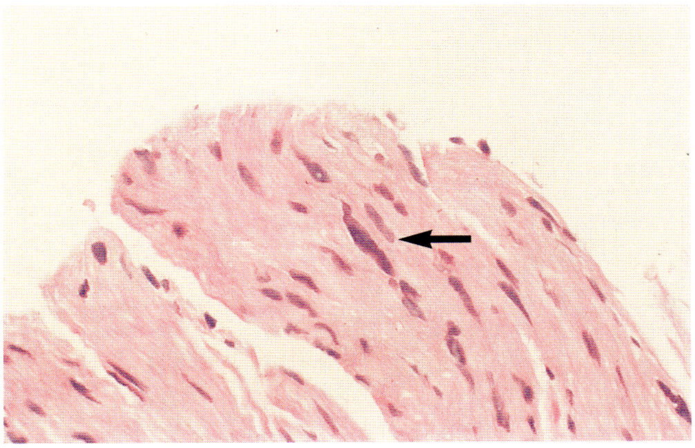

(c)

Figure 9.15 (c) High-power view of the tissue section shows the disorderly arrangement of the pleomorphic nuclei. The nuclear outlines are irregular. The appearances were thought to be compatible with a neurofibroma. However, the lesion recurred after excision. ×100.

RHABDOMYOSARCOMA OF THE HEAD AND NECK

This is a rare tumour in both adults and children, yet it is the commonest soft-tissue tumour of the head and neck region. Tumours are classified into three general histological categories: embryonal, alveolar and pleomorphic. However, depending on sampling, all these cell patterns may be seen within one tumour, which is then categorized according to the predominant component. Approximately 90% of childhood rhabdomyosarcomas are embryonal in type, whereas the alveolar type is commoner in adults. Poorly differentiated embryonal rhabdomyosarcomas are composed of small, round to oval nuclei surrounded by scanty cytoplasm. Cell borders are indistinct. These neoplasms are very well reviewed by Michaels (1990) and by Hellquist (1990). Cytological diagnosis of these lesions is useful where there is a known primary neoplasm with possible metastases or recurrence. Close clinicopathological correlation is always necessary before further treatment is undertaken. Definitive diagnosis of primary rhabdomyosarcomas is *not* possible on a cytological basis alone.

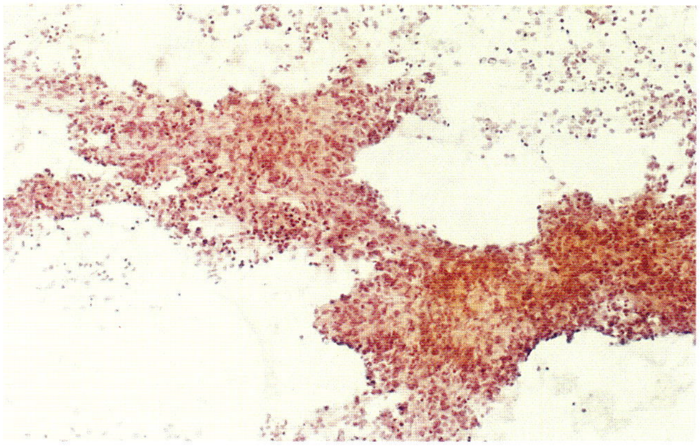

(a)

Figure 9.16 (a) FNA of a mass in the neck of an adult male. This low-power view shows a microbiopsy of tissue with a faint alveolar pattern, composed of very small, round cells with scattered loose-lying cells in the background. The cells show some arrangement in loose groups. Chromatin smearing (associated with small cell carcinoma) is *not* present. ×100.

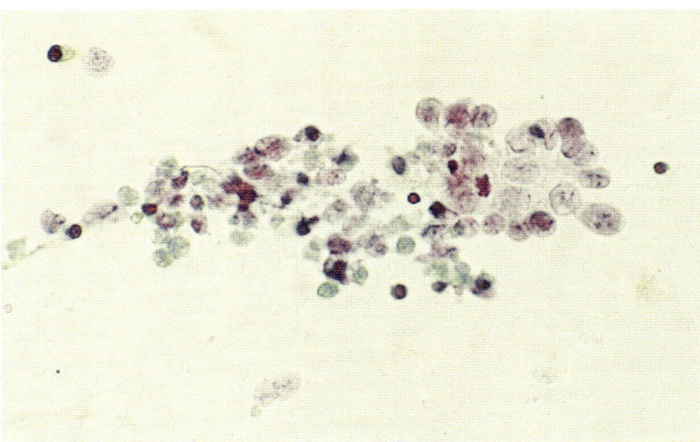

(b)

Figure 9.16 (b) High-power view of the specimen shown in (a), showing that the neoplastic cells are round, with a scanty rim of cytoplasm and round, pale nuclei with finely clumped chromatin giving it a speckled appearance. Many of the nuclei are intensely hyperchromatic: these have undergone apoptosis, a phenomenon associated with a rapidly growing neoplasm. No chromatin smearing is noted. These features are more suggestive of a sarcoma than an undifferentiated carcinoma. Close clinical correlation is suggested in such a case. ×400.

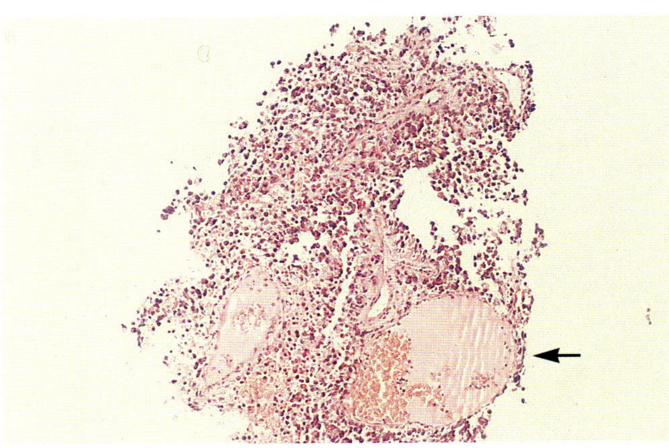

(c)

Figure 9.16 (c) Histological section of a Trucut biopsy of the mass discussed in (a) and (b) confirmed the presence of a neoplasm composed of small, round cells arranged rather haphazardly, although a faint alveolar pattern is discernible. Special stains confirmed that the neoplasm was a rhabdomyosarcoma, alveolar type. The eosinophilic focus of necrosis (arrow) may in fact be the needle-track resulting from the previous FNA. ×100.

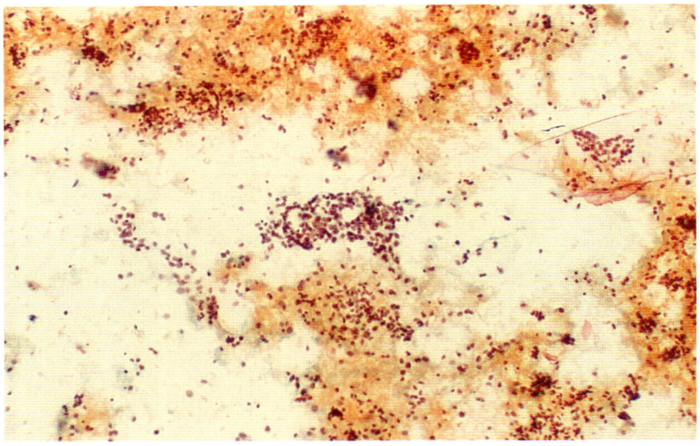

(a)

Figure 9.17 (a) FNA of a neck mass from a young patient with a previous history of alveolar rhabdomyosarcoma of the neck. This low-power view shows a cellular aspirate with numerous small undifferentiated cells, lying singly and in sheets. Chromatin smearing is *not* a feature. ×100.

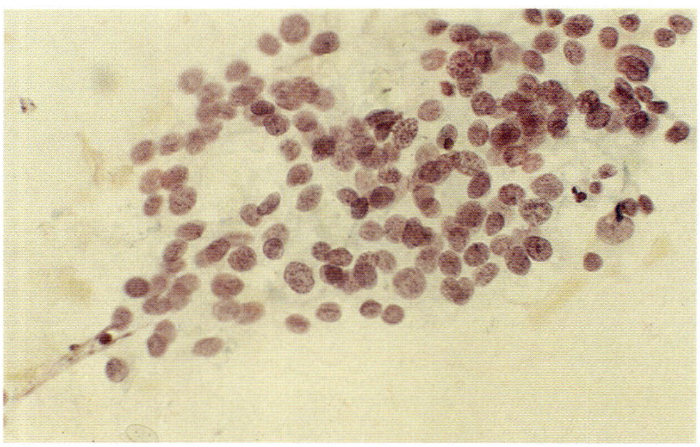

(b)

Figure 9.17 (b) High-power view of the specimen discussed in (a), showing small round cells with virtually no cytoplasm. The nuclei are fairly uniform and have regular outlines. The chromatin pattern is faintly speckled; *no* chromatin smearing is seen. Nucleoli are not present. Considering the clinical history, these appearances are more suggestive of a recurrent alveolar rhabdomyosarcoma rather than an undifferentiated carcinoma. ×400.

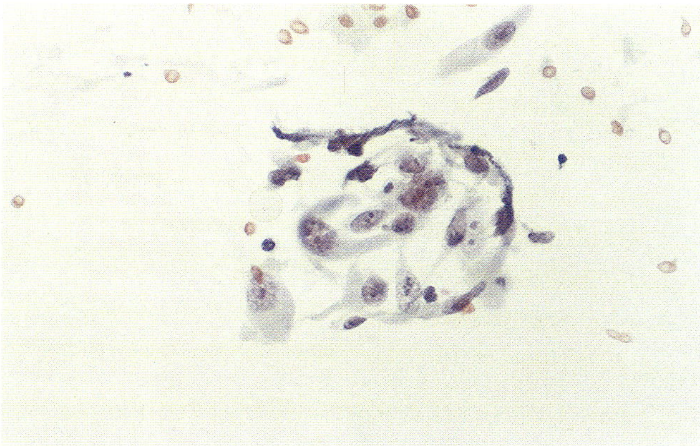

(a)

Figure 9.18 (a) FNA of a recurrent embryonal rhabdomyosarcoma of the neck in a child. Bizarre, very large pleomorphic cells are shown. They are mostly elongated, have plentiful cytoplasm, large pale nuclei with speckled chromatin and prominent rounded nucleoli. (The youth of the patient precludes a differential diagnosis of metastatic malignant melanoma.) ×100.

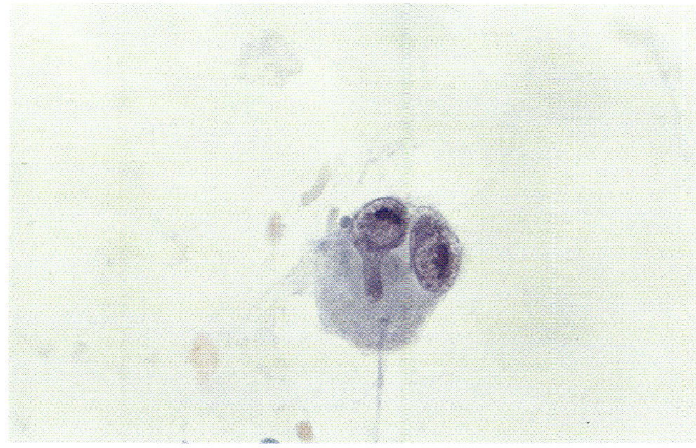

(b)

Figure 9.18 (b) High-power view of a single large bizarre cell from the same case: note the plentiful cytoplasm and the two large pale nuclei, which have eosinophilic, elongated and irregular macronucleoli. ×400.

(c)

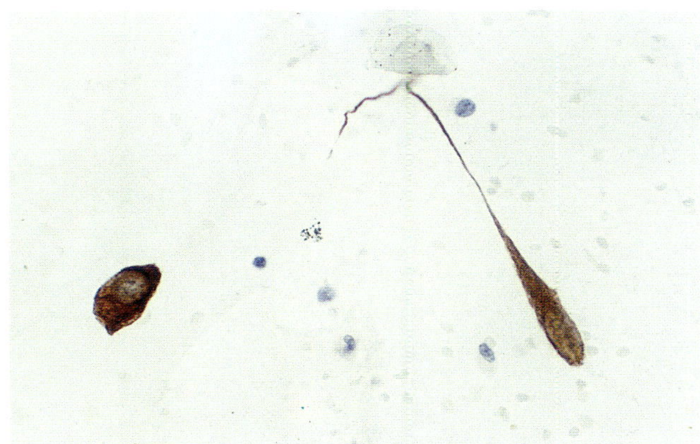

(d)

Figure 9.18 (c) and (d) One slide was reserved for immunocytochemical staining with Desmin. This was strikingly positive and emphasized the pleomorphism of the cells, as well as highlighting the extraordinarily long cytoplasmic tails, which are not obvious on the Papanicolaou stain. ×400.

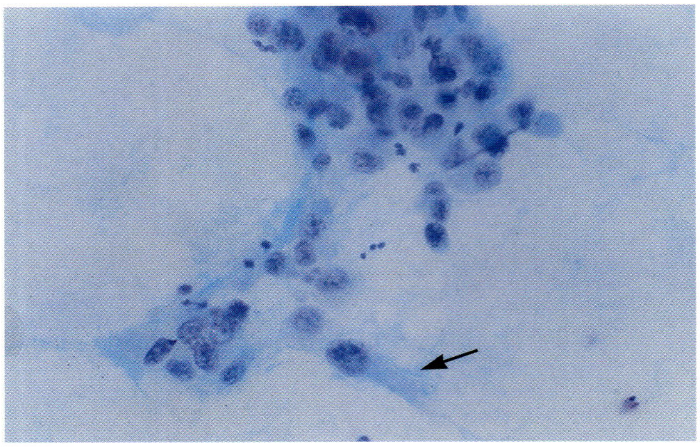

(a)

Figure 9.19 (a) FNA of a very large, rapidly enlarging mass in the neck of a child. The slides were referred from a clinic serving a rural area. No other clinical details were available. This flat sheet of pleomorphic cells shows degenerative changes which may be due to incorrect fixation or to biological degeneration. The cells have plentiful 'transparent' cytoplasm and pale nuclei with speckled chromatin and prominent nucleoli. Some cells have elongated cytoplasm (arrow). These appearances are suggestive of a sarcoma rather than a lymphoid neoplasm or a 'blast' cell neoplasm. There is no evidence of tuberculosis. ×400.

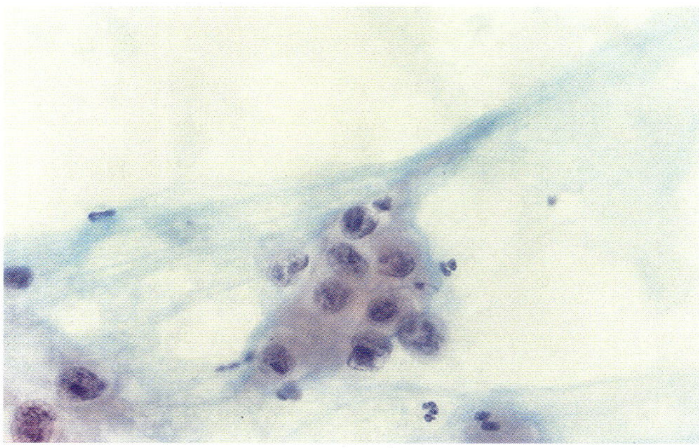

(b)

Figure 9.19 (b) Another group of cells from the case discussed in (a), showing pleomorphic cells with plentiful cytoplasm. The nuclei are pleomorphic, with disturbed chromatin and occasional prominent nucleoli. These appearances are suggestive of a sarcoma with uncertain differentiation. The patient did not return for further investigation. ×400.

REFERENCES AND FURTHER READING

General

Fagan, J. J., Learmonth, G. M., Garb, M. and Bowen, R. M. (1996) Nasopharyngeal lipoma: a rare clinicopathological entity. *J. Laryngol. Otol.* **110**: 275–276.

Fu, Y. S. and Perzin, K. H. (1977) Non-epithelial tumours of the nasal cavity, paranasal sinuses and nasopharynx. VIII Adipose tissue tumours (lipoma and liposarcoma). *Cancer* **40**: 1314–1317.

Grubb, C. (1988) *Diagnostic Cytopathology. A Text and Colour Atlas.* Edinburgh, Churchill Livingstone.

Hellquist, H. B. (1990) *Pathology of the Nose and Paranasal Sinuses.* London, Butterworths.

Learmonth, G. M. and Marks, R. K. (1988) Cytology in the diagnosis of bone lesions. *J. Bone Joint Surg (Br).* **70B**: 504. (Abstract.)

MacLeod, C. B. and Frable, W. J. (1993) Fine-needle aspirate biopsy of the salivary gland: problem cases. *Diagn. Cytopathol.* **9**: 216–225.

Michaels, L. (1990) *Atlas of Ear, Nose and Throat Pathology.* London, Kluwer.

Metastatic malignant melanoma

Koss, L. (1984) *Aspiration Biopsy. Cytological Interpretation and Histologic Basis.* New York, Igaku Shoin, pp. 254–257.

Kapila, K., Kharbanda, K. and Verma, K. (1991) Cytomorphology of metastatic melanoma. *Cytopathology* **2**: 229–237.

Perry, M. D., Gore, M., Siegler, H. F. and Johnston, W. W. (1986) Fine-needle aspiration of metastatic melanoma. *Acta Cytol.* **30**: 385–386.

Metastatic chondrosarcoma

Sanerkin, N. G. and Jeffree, G. M. *Cytology of Bone Tumours: a Colour Atlas with Text.* Bristol, Wright.

Stormby, N. and Akerman, M. (1973) Cytodiagnosis of bone lesions by means of fine-needle aspirate biopsy. *Acta Cytol.* **17**: 166–172.

Tsang, S. and Nguyen, G. K. (1988) Aspiration biopsy cytology of metastatic extra-skeletal myxoid chondrosarcoma and its differential diagnosis. *Acta Cytol.* **29**: 566–569.

Rhabdmyosarcoma

Ahkter, M., Ali, M. A., Bakry, M., Hug, M. and Sackey, K. (1992) FNA diagnosis of rhabdomyosarcoma. *Diagn. Cytopathol.* **8**: 465–474.

Perez-Guillermo, M., Bonmati-Limorte, C., Garcia-Rojo, B. and Hernandez-Gil, A. (1992) Infantile cutaneous rhabdomyosarcoma (Li–Fraumeni syndrome): cytological presentation of fine-needle aspiration. *Diagn. Cytopathol.* **8**: 621–626.

10 Cytopathology of the eye and orbit

Fine-needle aspirates of the orbital contents and periorbital tissue are occasionally received in the laboratory:

- for diagnosis of a clinically puzzling lesion
- for confirmation of an obvious lesion
- for confirmation of a metastatic lesion.

The commonest neoplasm occurring in the eye in childhood is retinoblastoma (Char and Miller, 1984). It has been reported that 90% of retinoblastomas are diagnosed before the age of 3 years (Buchino, 1991). However, in South Africa, children may present with advanced lesions at a later age. In hereditary cases, a second retinoblastoma in the contralateral eye may arise 12–18 months after the initial neoplasm. In clinically advanced cases spread can occur into the periorbital tissues, or recurrences may occur within the orbit post-inoculation. Metastases may also occur in the soft tissues of the face (Arora and Betharia, 1988).

The cell of origin is generally accepted to be the immature retinal cell. Although most retinoblastomas are composed of small, round, fairly uniform hyperchromatic cells with scanty or no cytoplasm, some tumours contain cells which show a marked degree of pleomorphism. These may be arranged in clusters with or without rosette formation. In high-grade lesions there may be apoptosis (individual cell necrosis). This may also be seen after irradiation. Mitotic figures may be present. Pigmented retinal cells or macrophages containing melanin pigment may be present in the background.

Malignant melanoma may occur primarily in the retina in adults; however, it may also be metastatic to the eye, so when considering a diagnosis of melanoma the clinical history should always be sought, with particular reference to previously removed pigmented lesions.

Rarely lacrimal gland tumours present a puzzling clinical problem. Differentiation from a metastatic neoplasm is necessary. Cystic masses in the orbital region are in rare instances parasitic hydatid cysts due to *Echinococcus granulosus*. Hydatid disease is endemic in the Cape Province in South Africa (Learmonth and Murray, 1990).

RETINOBLASTOMA

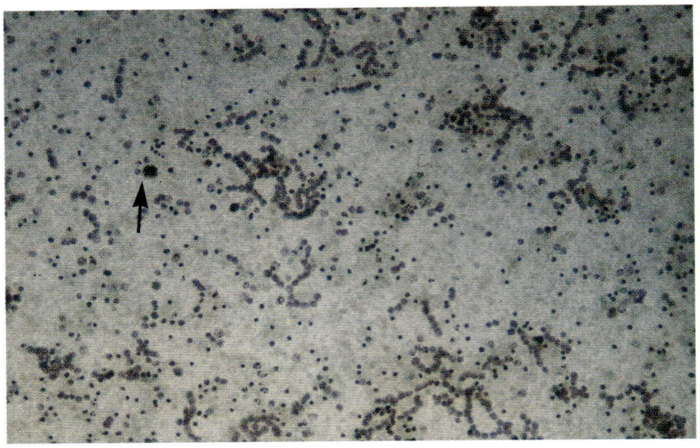

Figure 10.1 This is a sample of vitreous fluid from the eye of an 8-year-old boy with familial retinoblastoma. The small, uniform round cells have a curious arrangement of chains and circles which has been likened to 'Chinese writing'. Scattered, loose-lying, paler, degenerate cells are noted in the background, together with pigmented cells (arrow). ×100.

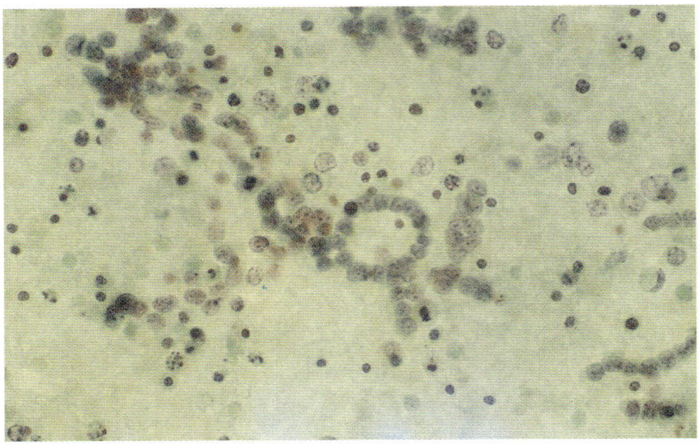

Figure 10.2 Cells from a retinoblastoma with a circular arrangement (a rosette); the nuclei have disturbed chromatin and show mild pleomorphism. Smaller, darker, single-lying lymphoid cells are seen in the background. ×400.

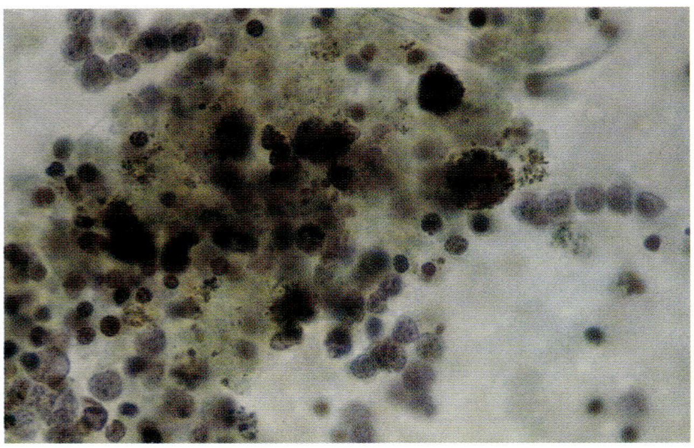

Figure 10.3 Melanin pigment in cells from the same case as shown in Figures 10.1 and 10.2. These pigmented cells occur in the normal retina and are seen here interspersed with tumour cells. ×400.

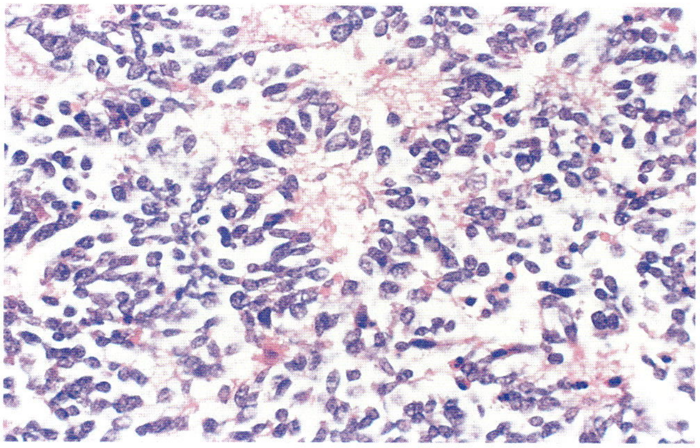

Figure 10.4 Histology of retinoblastoma, showing a tumour consisting of small, round, fairly uniform cells forming a rosette with central eosinophilic amorphous material. H&E ×100.

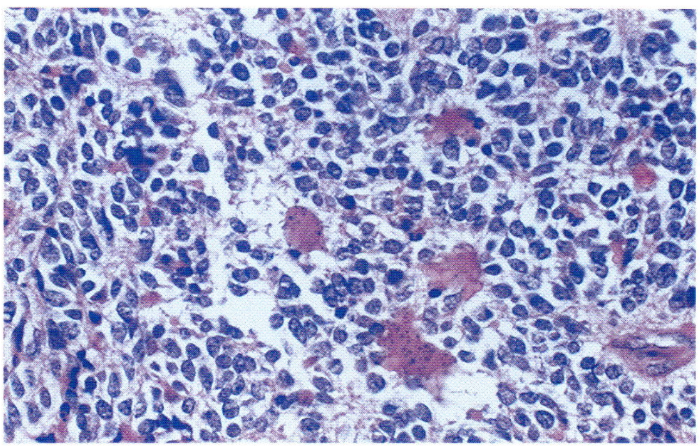

Figure 10.5 Necrosis, manifested by small eosinophilic foci containing fragments of basophilic nuclear material, is seen in this view of a retinoblastoma. H&E ×100.

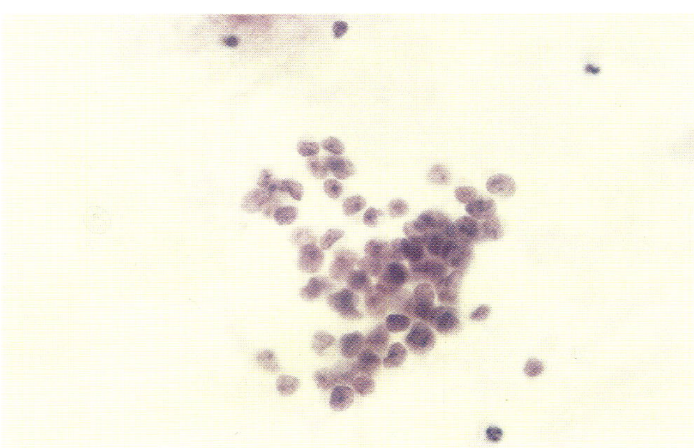

Figure 10.6 Cytological preparation from the eye of a 5-year-old girl thought to have postinfectious scarring with resultant glaucoma. Numerous clusters of small, uniform undifferentiated hyperchromatic cells with prominent nucleoli and irregular nuclear outlines are present, compatible with a retinoblastoma. These cells can be differentiated from lymphoid cells as they are in clusters and are not single-lying like lymphoid cells. ×400.

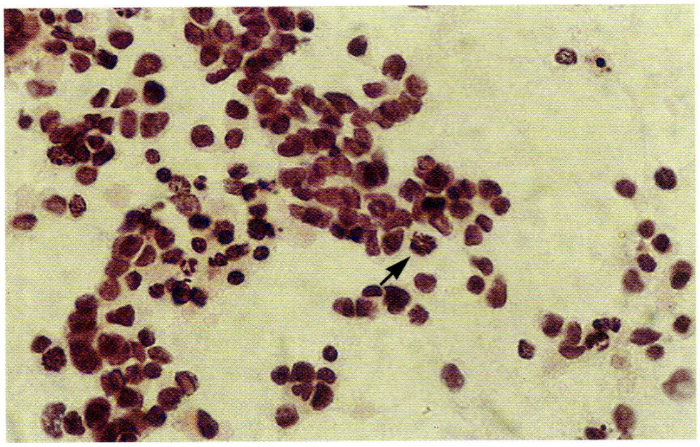

Figure 10.7 These cells are from the right eye of a child with a history of a retinoblastoma in the left eye 1 year previously. The hyperchromatic undifferentiated cells show some pleomorphism and individual cell necrosis. Mitotic figures are present (arrows). ×400.

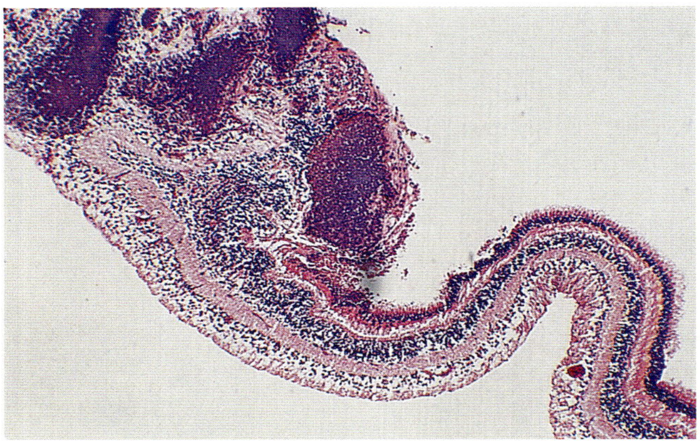

Figure 10.8 Histological appearance of the retinoblastoma shown in Figure 10.7 shows nests of small hyperchromatic tumour cells within the retina. H&E ×100.

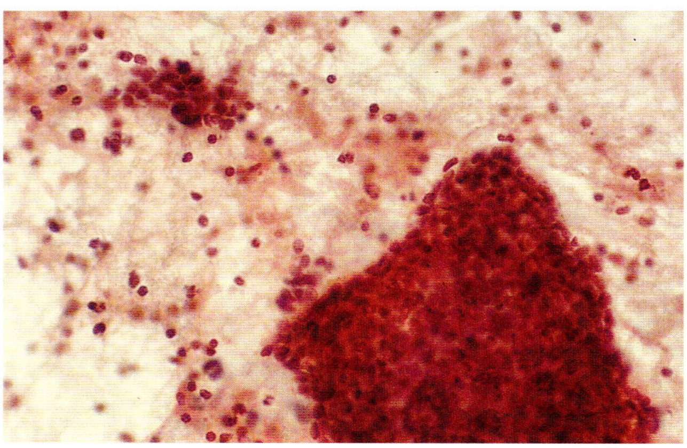

Figure 10.9 FNA of a nodule in the enucleated left orbit 2 years after surgery and radiotherapy shows microbiopsy of residual retinoblastoma as well as a cluster of tumour cells and single-lying cells. ×100.

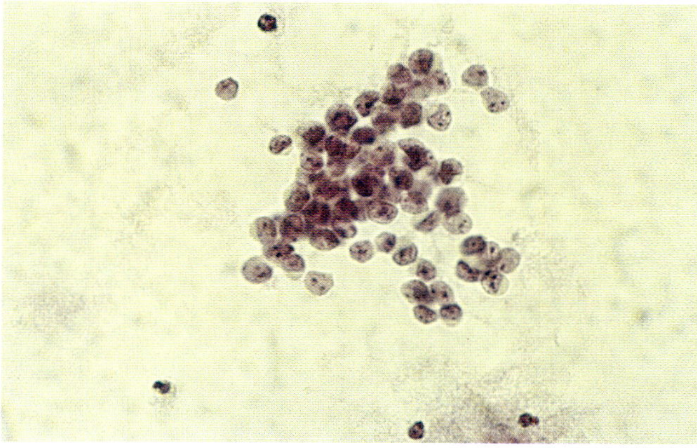

Figure 10.10 Well-preserved cells from a retinoblastoma in fluid from the anterior chamber showing the fairly uniform size and shape of the nucleoli; nuclei are prominent. ×400.

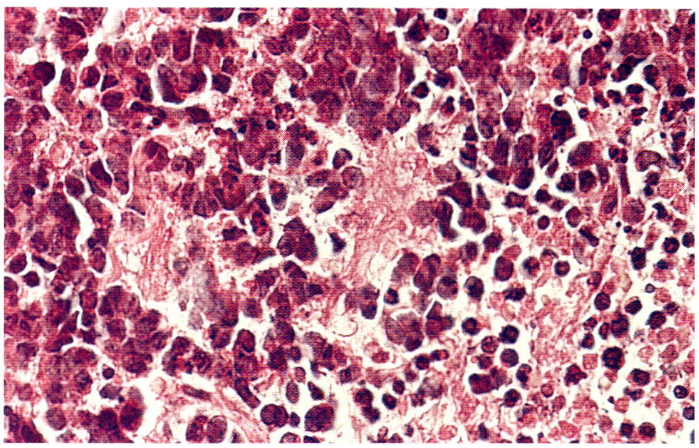

Figure 10.11 This histological section of a retinoblastoma shows the pleomorphic nature of the tumour cells seen in Figure 10.10. H&E ×400.

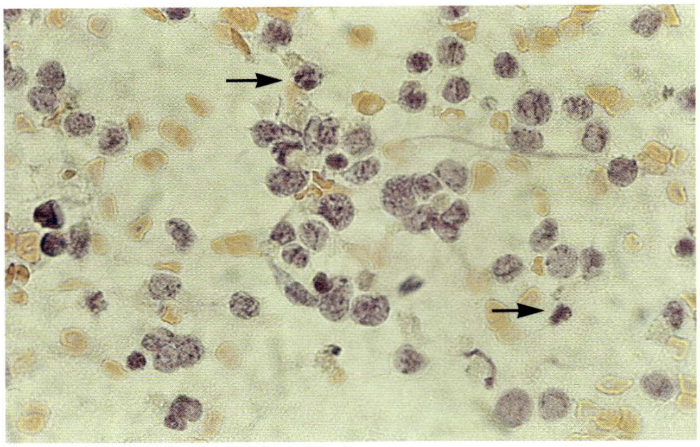

Figure 10.12 FNA of a recurrent nodule in the orbit of a child 1 year post-enucleation. The loose-lying cells show marked pleomorphism with pale, markedly disturbed chromatin. Individual cell necrosis (apoptosis) is noted (arrow). The latter feature is seen in rapidly growing tumours and in patients who have had radiotherapy or chemotherapy. ×400.

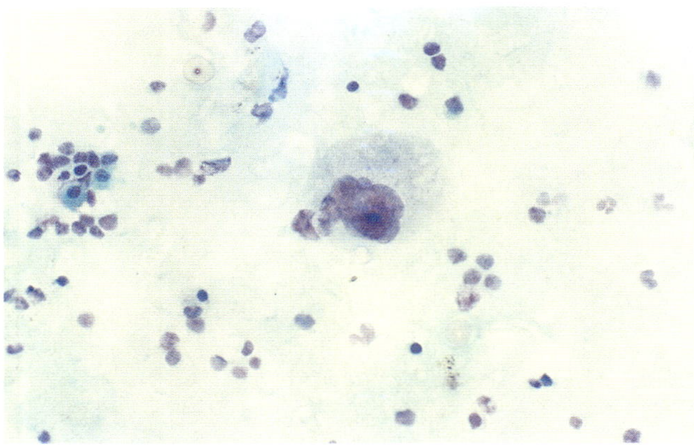

Figure 10.13 This large anaplastic cell was seen in an FNA specimen from a recurrent retinoblastoma. This cytopathic effect may be due to therapy. ×400.

METASTATIC MALIGNANT MELANOMA

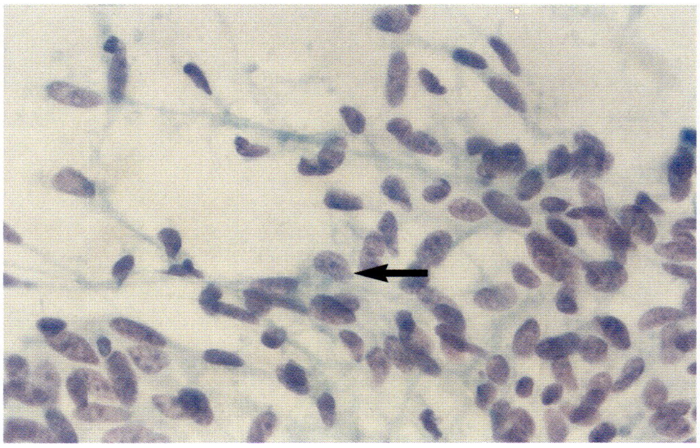

Figure 10.14 FNA of a nodule on the inner canthus of the eye in a patient who had a primary malignant melanoma of the nasal columnella. The specimen was very cellular, with loosely cohesive, small, spindle-shaped cells with disturbed chromatin and some prominent nucleoli. Although no melanin pigment is present the features are compatible with a metastatic spindle-cell melanoma (see Chapter 2).

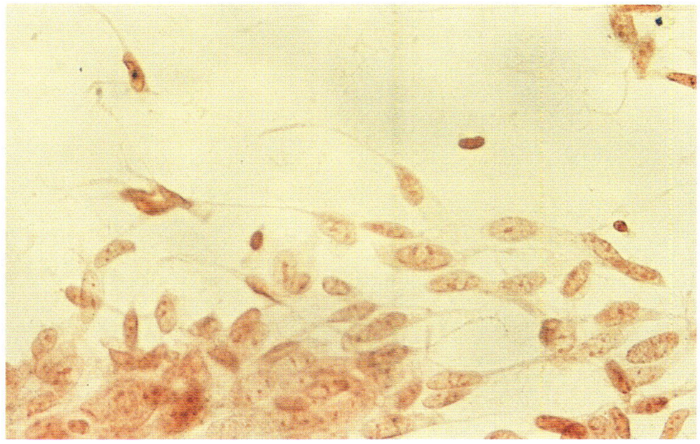

Figure 10.15 Immunostaining of a destained slide from the specimen shown in Figure 10.14 shows that the stain for melanin is positive. Note the long cytoplasmic tails which, together with the nuclear details, are highlighted with this stain. ×400.

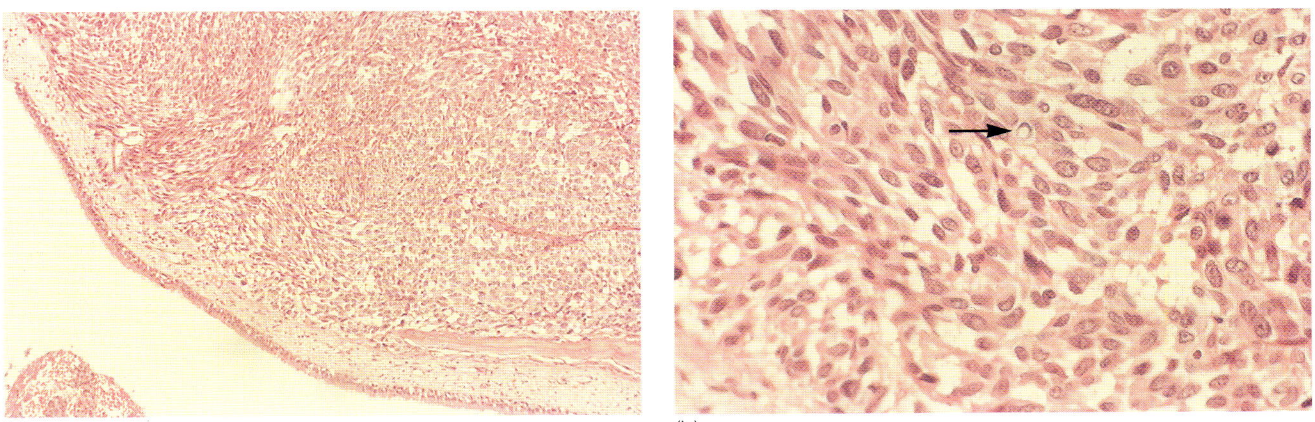

(a) (b)

Figure 10.16 (a) and (b) Histology of the primary malignant melanoma of the nose discussed in Figure 10.14 showing the spindle-cell nature of the primary neoplasm, which is situated beneath the respiratory epithelium. An intranuclear vacuole is noted (arrow). ×100 and ×400.

LACRIMAL GLAND TUMOUR

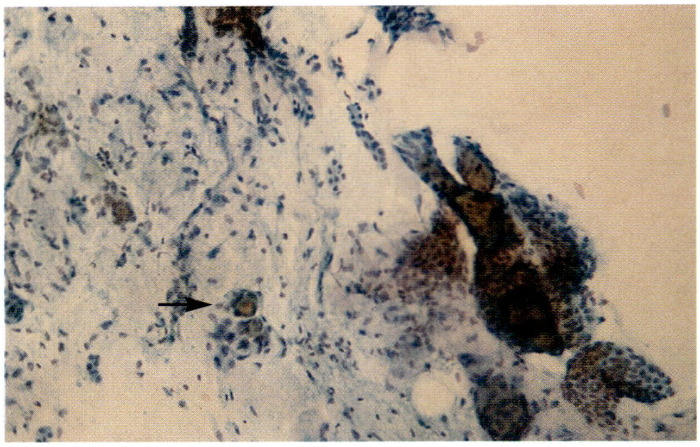

Figure 10.17 FNA of a mass in the lacrimal gland in an adult. Microbiopsies of loose stromal tissue and sheets and groups of fairly uniform epithelial cells are present, consistent with a pleomorphic adenoma. A squamous metaplastic cell is present in the centre of one nest of cells (arrow). The epithelial cells are tightly cohesive and have a trabecular or ductal arrangement. ×100.

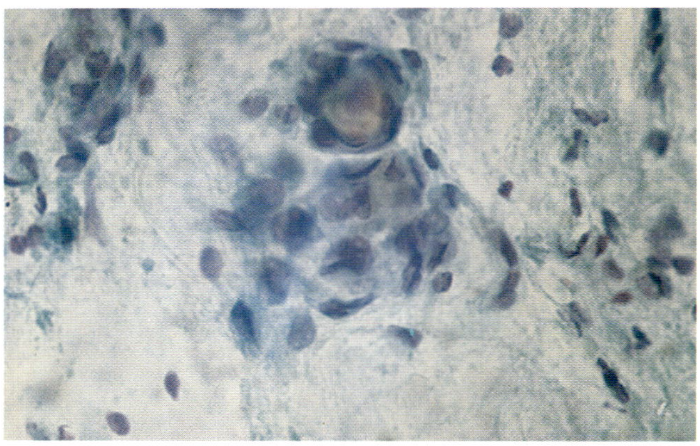

Figure 10.18 High-power view of the epithelial cells seen in Figure 10.17 from a pleomorphic adenoma of the lacrimal gland. The central cell shows the features of squamous metaplasia. The cells look quite pleomorphic and could be mistaken for a low-grade malignancy, if the entire clinical and cytological picture (especially the background of loose stroma) is not taken into account. ×400.

PARASITIC CYST

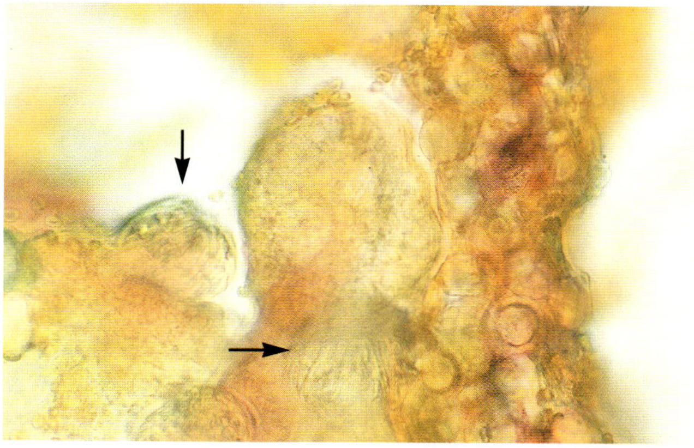

Figure 10.19 FNA of an orbital cyst in a young woman. Cytological preparation of the clear fluid which was aspirated showed spherules of calcific material and clusters of refractile hook-like structures. This material was originally dismissed as 'debris' by the cytotechnician; however, awareness of parasitic disease in the Cape Province (Learmonth and Murray, 1990) led to recognition of this material as the contents of a hydatid cyst caused by the tapeworm *Echinococcus granulosus* (see Chapter 1). The calcific material is known as 'hydatid sand' and the tight cluster of hooklets is termed a rostellum. The recognition of parasitic cysts on cytology is very important from a clinical point of view, as the surgeon needs to be alerted to the danger of an anaphylactic reaction in the patient during the removal of the cyst (Knudsen *et al.*, 1988). ×400.

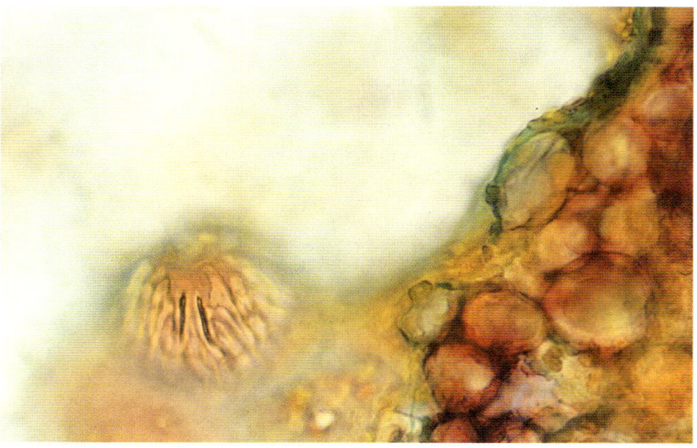

Figure 10.20 Another view of a rostellum of hooklets with adjacent clusters of hydatid sand. ×400.

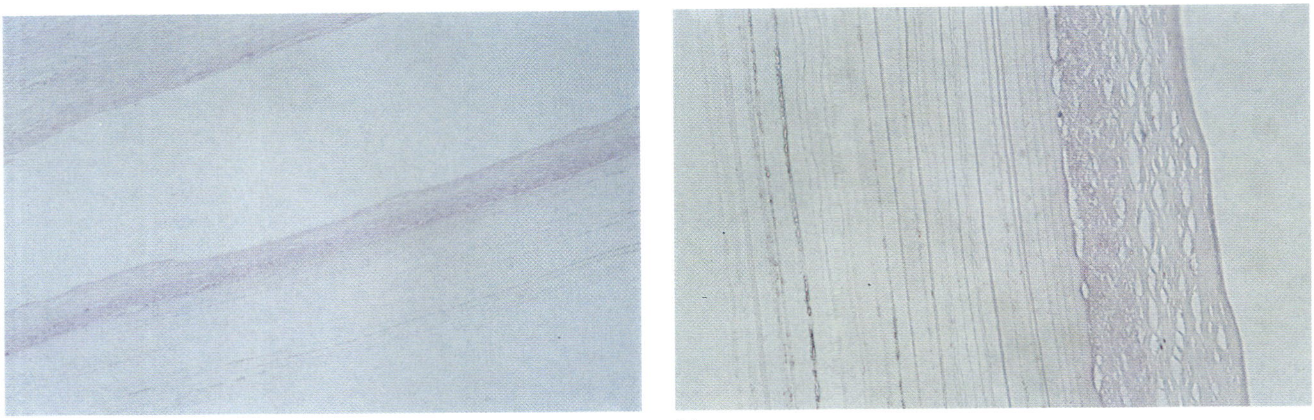

(a) (b)

Figure 10.21 (a) and (b) Histological section of the fragile cyst wall: it is pale-staining and, on close examination, may show characteristic laminations. The cyst contents are usually lost during histological processing; however, cytospin preparations of the formalin fluid from the specimen bottle will usually reveal parts of the parasite. ×100.

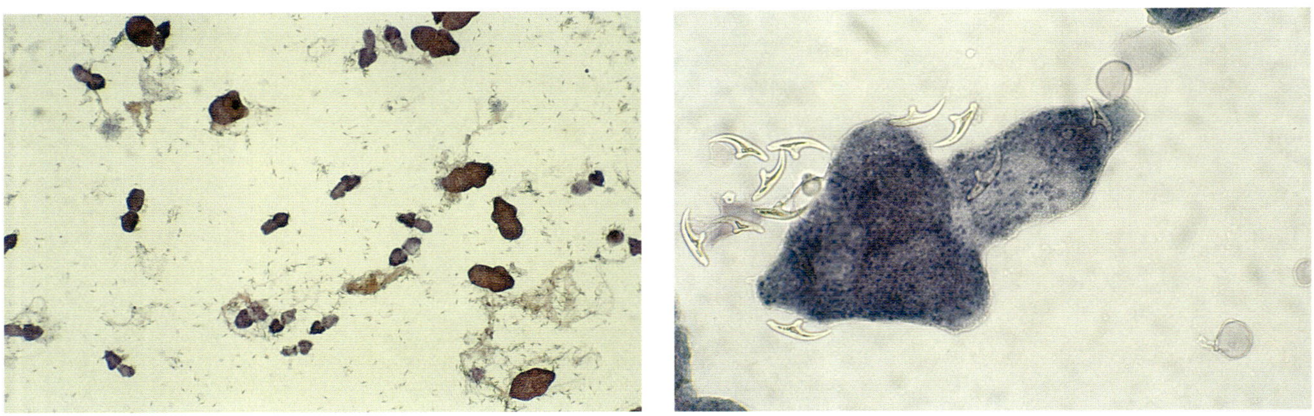

(a) (b)

Figure 10.22 (a) and (b) Cytospin preparation of formalin fluid containing the specimen seen in Figures 10.19–10.22 shows the head of the tapeworm with attached body segments. The lateral suckers are well shown in the high-power view. The refractile nature of the sharply outlined hooklets creates quite a dramatic photomicrograph. ×400.

REFERENCES AND FURTHER READING

Retinoblastoma

Arora, R. and Betharia, S. M. (1988) Fine-needle aspiration of metastatic retinoblastoma. *Acta Cytol.* **32**: 428–429.

Buchino, J. J. (1991) *Cytopathology in Paediatrics*. Monographs in Clinical Cytology, Vol. 13. Berlin, Karger.

Char, D. H. and Miller, T. R. (1984) Fine-needle biopsy in retinoblastoma. *Am. J. Ophthalmol.* **97**: 686–690.

Howell, L. P. (1992) The cytology of paediatric masses: a differential diagnostic approach. *Diagn. Cytol.* **8**: 107–115.

Metastatic malignant melanoma

Travis, W. D. and Wold, L. E. (1987) Immunoperoxidase staining of fine-needle aspiration specimens previously stained by the Papanicolaou method. *Acta Cytol.* **31**: 517–520.

Echinococcus granulosus – hydatid cyst

Learmonth, G. and Murray, M. (1990) Helminths and protozoa as an incidental finding in cytology specimens. *Cytopathology* **1**: 163–170.

Knudsen, C., Marks, R. K. and Learmonth, G. (1988) Orthopaedic hydatid disease. (Abstract.) *J. Bone Joint Surg (Br)*. **70B**: 504.

Sodhani, P., Das, D. K., Nalarajan, R. and Khan, I. U. (1996) Cytological diagnosis of orbital echinococcosis. *Diagn. Cytopathol.* **15**: 52–53.

Index